Electric Arc Furnace Steelmaking

Electric Arc Furnace Steelmaking

Miroslaw Karbowniczek

CRC Press is an imprint of the
Taylor & Francis Group, an **informa** business

First edition published 2022
by CRC Press
6000 Broken Sound Parkway NW, Suite 300, Boca Raton, FL 33487-2742

and by CRC Press
2 Park Square, Milton Park, Abingdon, Oxon, OX14 4RN

CRC Press is an imprint of Taylor & Francis Group, LLC

ISBN: 9780367673475 (hbk)
ISBN: 9780367673482 (pbk)
ISBN: 9781003130949 (ebk)

DOI: 10.1201/9781003130949

Typeset in Times
by codeMantra

Contents

Preface

Issues related to steelmaking in electric furnaces rarely become a subject of monographs. This primarily stems from the fast technological development in this area. The term "steel electrometallurgy", meaning the area of technology encompassing steelmaking wherein electricity is the source of heat, emerged at the beginning of the 20th century.

Steel production in electric arc furnaces (EAFs) is becoming increasingly important, which is proven by the global steel output, which reached almost 500 million metric tons in 2019 (an approximately 30% share in the overall steel production). Due to the dynamic development of technology and science, there is a need to continuously learn and understand the new, in-depth knowledge in the area concerning the rudiments of steelmaking processes in EAFs. The knowledge of both the construction of this device and its technical capabilities enables this device to be optimally, cost-effectively, and environmentally friendly used in industrial conditions, and it also enables the process optimization to be further developed and improved.

This book presents a systematic and exhaustive discussion of the current knowledge of metallurgical processes carried out in EAFs. It describes the principles of the construction of EAFs as well as the applied design solutions and their operation (along with ancillary equipment). It also presents the modern practices of making various steel grades in detail, along with an explanation of the theoretical rudiments of the occurring processes and chemical reactions.

The book comprises 10 chapters, of which Chapters 1–7 are concerned with development of the design and principles of operation of EAFs. Chapters 8 and 9 are concerned with the charge materials and steel (Fe-C alloys) melting practice, respectively. Chapters 2–5 present the construction and principles of the operation of modern EAFs with various capacities, along with the power supply systems and auxiliary equipment to support the steelmaking process. Chapter 6 offers an interesting description of the final component of the electric arc power supply system, which are graphite electrodes. Chapter 7 is concerned with the issues of environmental protection against the harmful impact of some effects accompanying the steelmaking process, whereas Chapters 8 and 9 are devoted to the steelmaking practice in an EAF taking into account the available modern solutions. Chapter 10 presents the principles of preparation of the mass and heat balances of the steelmaking process, including calculation examples. This monograph contains theoretical analyses and results of laboratory, model, and industrial tests. It is intended for students of bachelor's, master's, and doctoral studies as well as for engineers working in the steelmaking, nonferrous metals industry or in an iron and steel foundry. The book primarily focuses on the practical, industrial aspects of EAF operation management, including environmental issues.

This book was created on the basis of many years of experience in preparing and delivering lectures on steel electrometallurgy for the students of the AGH University of Science and Technology, and in working with doctoral students. The possibilities of the current observation of introducing new design and technical solutions in the operation of EAFs stemmed from the completion of numerous projects, many years of collaboration with companies designing steelmaking EAFs, and developing the practices applied in them as well as with steel plants. The collaboration with SMS Group (formerly SMS Siemag), which is a global leader in the design and implementation of increasingly advanced new solutions in the construction and use of EAFs, deserves special emphasis. I would like to thank the employees of the SMS Group, especially Dr. Jan Reichel for permission to include some of the drawings. Similar thanks are due to the employees of the Zakłady Magnezytowe "Ropczyce", Poland, for their long-term cooperation in the field of refractory materials used in electric furnaces and for giving consent for the use of several drawings in the manuscript.

This book is prepared on the basis of the book *Stalowniczy piec lukowy* that was published in Polish in 2015 by Wydawnictwa AGH (ISBN: 978-83-7464-807-3).

Author

Professor Miroslaw Karbowniczek, PhD, is a professor at AGH University of Science and Technology (AGH-UST) in Krakow, Poland. He was the Dean of the Faculty of Materials Engineering and Industrial Computers Science (2008–2012) and a Vice-Rector of AGH-UST (2012–2020). His academic interests focus on metallurgy, especially the metallurgy of steel, electric arc furnace metallurgy, ladle metallurgy, automatics, control, and modeling in this area.

He has authored or coauthored more than 220 scientific papers, including two original monographs, one original textbook, and ten patents (including one registered in most countries worldwide). He has on many occasions presented the findings of his research at international scientific conferences. The international academic community has expressed their recognition by entrusting him with organizing and chairing the 9th European Electric Steelmaking Conference, which he organized in Krakow in 2008.

1 Introduction

1.1 CHARACTERISTICS OF STEEL ELECTROMETALLURGY

Metallurgy is an area of science and practice covering manufacturing, processing, and research on metal properties. The objective of the work in the area of metallurgy is to process the metal ores that are present in the earth's crust and to obtain the final product in the form of finished products that are useful in people's lives. It includes the extracting processes, metal casting, metal forming, and creating appropriate product properties. It is often widely considered equivalent to steelmaking, but steelmaking has a narrower meaning; it covers the practical, industrial aspect of obtaining metal products, and it is particularly concerned with the so-called extraction metallurgy. Metal extraction processes obtain a metal ingot from metal ores for further processing. Electrometallurgy is an area of extraction metallurgy. This is the part of metallurgy theory and practice that deals with manufacturing metals and their alloys in the furnaces wherein electrical energy is the source of heat. With respect to the production of ferrous alloys, it is called steel electrometallurgy. The use of electrical energy as a heat source in the steelmaking process has a number of advantages; the capability to control the temperature is one of the most important.

The electric arc furnace (EAF) is the basic industrial unit for steel production using electrical energy. Here, electrical energy is converted into heat energy in an electric arc. Nowadays, over 1.8 billion metric tons of steel is produced worldwide annually, including almost 30% in EAFs. EAFs are particularly useful for steel production using steel scrap as the charge. Practically all steel scrap that is generated as a result of human activity is converted in metallurgical processes into useful steel.

1.2 HISTORICAL DEVELOPMENT OF ELECTRIC FURNACE DESIGN

The development of EAF design is inseparably related to the development of knowledge of electrical engineering, and to the progress in science and technology of the electrical industry as well as the techniques of control, automation, and IT. The effect of the electric arc was discovered in 1809 by the English scientist Humphry Davy, when he experimented with primary batteries [1]. By and large, it has been since that time that attempts began to use the electric arc for melting metals. The first laboratory EAF was built in the mid-19th century; at that time, the first patents for melting metals with the electric arc were also obtained. William Siemens was the forerunner, who was the first to design laboratory EAFs: direct heating furnace in 1878 (Figure 1.1) and a furnace with a conductive hearth and direct charge heating in 1879 (Figure 1.2). In these furnaces, the automatic control of the arc current was applied

DOI: 10.1201/9781003130949-1

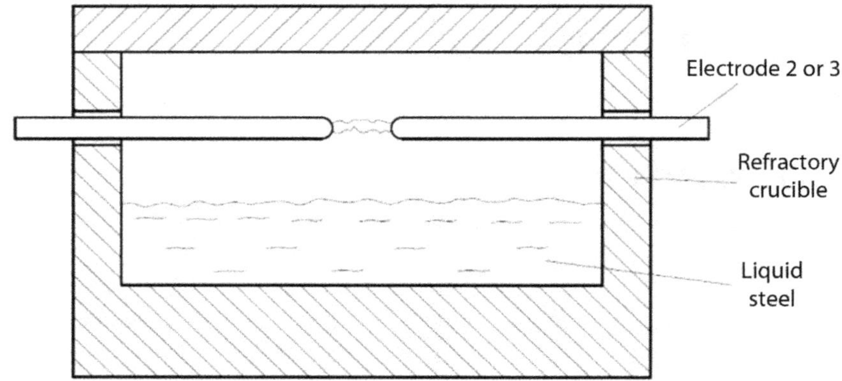

FIGURE 1.1 Diagram of indirect electric arc heating [2]. (Adopted from Charles, 1985.)

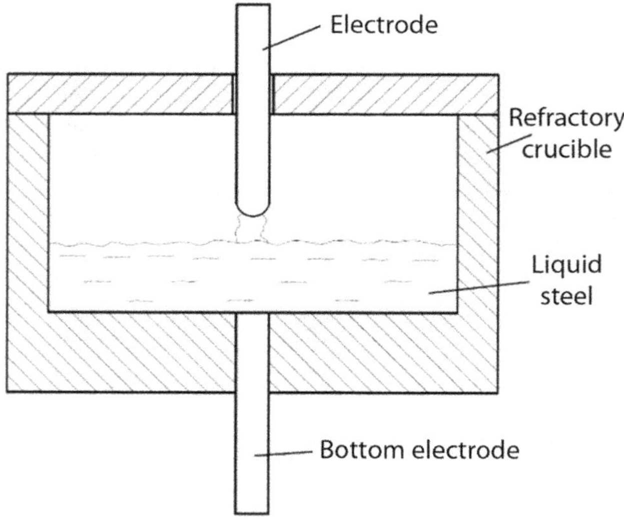

FIGURE 1.2 Diagram of direct electric arc heating [2]. (Adopted from Charles, 1985.)

for the first time by changing the supply electrode position with an electromagnetic controller.

In 1888, there was a breakthrough in the development of steelmaking EAF design when Paul Héroult patented an EAF with the direct heating of the metal bath (Figure 1.3). In this furnace with the direct heating of the metal charge, the current flows in a closed circuit: electrode – slag – metal – slag – electrode. After applying many improvements and modifications, as operating experience was gained, theoretical and applied electrical engineering developed, and in 1899, this furnace became the basis for designing the first prototype of industrial EAF, of which the basic design

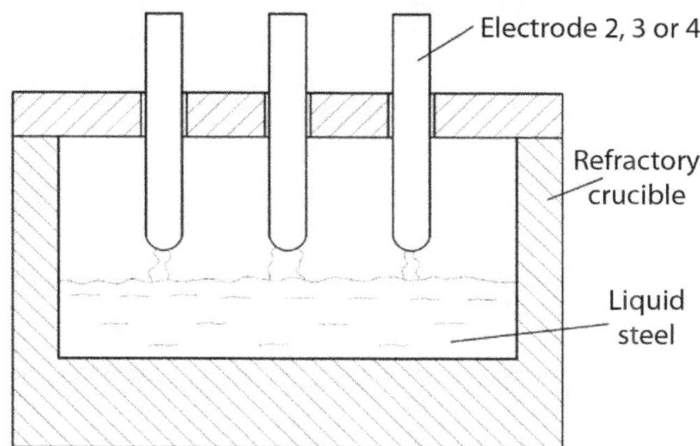

FIGURE 1.3 Diagram of the process with direct electric arc heating [2]. (Adopted from Charles, 1985.)

principles are still being applied. Steel production on an industrial scale in an EAF of this type started in 1907.

The original design of the EAF by P. Héroult had a disadvantage: a nonuniform distribution of temperature in the metal bath arising from an insufficient specific power of the furnace transformer and an excessively low electric arc current. Therefore, the charge melting was slow, and the heat losses accompanying this process were high. Consequently, the unit electricity consumption was also high. The basic design advantage of the direct EAF, which was the possibility of high power concentration and thereby the acceleration of the charge melting process, was not utilized. Therefore, induction core furnaces and indirect EAFs were competitive for direct EAFs.

The industrial implementation of direct EAFs in the steel metallurgy was stopped because of an incorrect focus on new design solutions. Based on the assumption that the operation of this furnace can be substantially improved by creating increased convection resulting from the electric current impact, design solutions enabling the electrical current to flow through the whole depth of the metal bath and the furnace hearth were sought. Between 1900 and 1915, a series of design solutions of direct arc furnaces were introduced with a water-cooled electrode embedded in the hearth, that is, the so-called conductive hearth furnaces. These furnaces (Gironde and Nathusius furnaces), in terms of design, were more complicated, and they did not confirm their advantages in industrial operation. This was caused by insufficient strength of the hearth. Therefore, they were gradually eliminated from production and replaced with improved furnaces with a nonconductive hearth. The implementation of threaded carbon electrodes between 1910 and 1911 for production, followed by graphite electrodes, was significant for the development of steelmaking arc furnaces. These electrodes, operating continuously (possibility of their extension), stabilized the operating conditions of EAFs, which improved their performance.

In the 1920s, Andrea and Ricke developed circular charts and the electrical characteristics of arc furnace operation. The analysis of these charts indicated that the fundamental design fault of arc furnaces was due to the application of excessively low operating voltages (90–130 V). Increasing these voltages to the range of 180–230 V allowed the unit power and furnace efficiency to increase significantly as well as to improve their thermal efficiency.

Attempts were also made to apply electrical energy for metallurgical production in furnace designs other than the classic arc furnaces. In 1905, M. Bolton, in a Siemens plant, built the first prototype of a vacuum arc furnace, operating in an argon atmosphere or at a negative pressure. Vacuum arc furnaces were only applied commercially in the 1950s while simultaneously initiating a whole group of various designs of furnaces for the secondary refining of steel. This group included induction furnaces, electron-beam furnaces, furnaces for electro-slag steel refining, and plasma furnaces.

The first attempts to use the electromagnetic induction effect for melting metals were made at the beginning of the 20th century. However, already in 1887, D. Ferranti patented a low-frequency induction core furnace. In 1890, E.A. Colbe applied for a patent for a vacuum core induction furnace. The first industrial core induction furnace was built by Kjellin in Sweden in 1908. However, core induction furnaces were not applied in steel electrometallurgy due to an unfavorable shape of the working space. They are applied in nonferrous metallurgy.

During 1905–1907, A.N. Lodygin designed – and in 1908 published – a description of the principles of operation and design of the high-frequency coreless induction furnace. At the same time, French company Schneider–Creusot obtained a patent for the coreless induction furnace. The first trials of melting zinc with a high-frequency electric current were conducted by the Lorentz Company in 1912–1913. A coreless induction furnace for melting steel in the industrial scale was built in the US by E. Northrup in 1916.

Initially, coreless induction furnaces were not applied commercially due to severe difficulties related to the generation of high-frequency electric current. Only the development of the design of machine generators for high-frequency current led to the construction of the first industrial coreless induction furnace for steel melting in 1930. In the subsequent years, the development of coreless induction furnace design was fast, and it was accompanied by a constant increase in their capacity. At present, open and vacuum high-frequency induction furnaces are applied to make top quality steel and special alloys. The capacity of these furnaces achieves approximately 65 t. As the technology has developed, thyristor frequency converters have been applied, which has substantially simplified their design and allowed the current frequency to be controlled.

1.3 DEVELOPMENT OF THE DESIGN OF MODERN ELECTRIC ARC FURNACES

So far, we can distinguish four generations of steelmaking EAFs, with reference to their design and the practices applied [3]. The unit power of the installed furnace transformer and the related productivity can be adopted as the basic criteria for division. The first-generation furnaces in use until the 1960s had transformers up to 250 kVA/t of furnace capacity, and their productivities reached 15–20 t/h. Works by W.E. Schwabe on the design of the secondary circuit and high-power technology published in the 1960s contributed to the construction of furnaces with the transformer unit capacity from 250 to 450 kVA/t. It enabled their productivity to increase from 25 to 40 t/h. A dozen or so years later, the third-generation furnaces with the transformer capacity from 450 to 700 kVA/t were implemented, and their productivity was 50–80 t/h. For about 30 years, most furnaces that were built have been fitted with a transformer with a unit power over 700 kVA/t (often above 1,000 kVA/t), and the obtained capacity reaches above 100 t/h. The first-generation furnaces were called RP (regular power), the second – HP (high power), the third – UHP (ultra-high power), and the last – SUHP (super ultra-high power). The development of transformer power also involved an increase in the furnace capacity. The first furnaces had capacities in the range of 5–25 t, whereas the capacity of the ones built today reaches 400 t.

The first-generation furnaces had the hearth and walls lined with magnesite or dolomite refractories. The roof was first built from silica materials and later from alumina materials with an increasing Al_2O_3 content. The metal bath was oxidized with iron ore or scale, and the meltdown period time was about 3 h, while the whole tap-to-tap time, depending on the practice applied and steel grades manufactured, was between 4 and 8 h.

The first arc furnaces featured the relatively fast and uneven consumption of the refractory lining. Research conducted on the heat transfer inside the furnace has led to the development of relationships describing the furnace supply conditions and lining consumption. Thanks to the reduction of the electrode circle diameter and a decrease in the arc voltage, and an increase in the current intensity, it has become possible to operate at higher powers, while extending the refractory lining life. It caused the necessity of using graphite electrodes with a low resistivity, capable of conducting high-intensity current and the high-current circuit with water-cooled wires designed and arranged to minimize asymmetry. Environmental protection requirements have forced the users to employ the fourth hole in the roof to capture off-gases. At that time, the notion of an operation time utilization indicator, as the quotient of time of drawing power and the tap-to-tap time, was also introduced. Therefore, organizational changes were made to reduce breaks for charge loading, fettling, etc. As previous furnaces were almost solely used for the production of highly alloyed steels, the second-generation furnaces made it possible to economically manufacture low alloy and even carbon steels.

Although the operation of EAFs with a low arc voltage and a high current intensity caused an increase in the refractory lining life, it simultaneously caused an increased consumption of more and more expensive graphite electrodes. Further searching for the possibilities of increasing the furnace performance has led to designing a new, third generation of furnaces. The research work conducted and design changes of EAFs made at that time have led to such an increase in their productivity that steelmaking in an EAF has become competitive for the existing BOF shops, even for carbon and low-alloy steels. It was possible primarily as a result of progress in the work on the design of power supply systems and the optimization of operating parameters, which has led to a reduction of the unit manufacturing costs. In the initial solutions, the power supply system worked at arc current intensities below their values corresponding to the maximum arc power and the maximum active power. This caused an incomplete utilization of the arc power and an extension of the tap-to-tap time and an increase in heat losses. At the same time, the analysis of heat transfer conditions inside the furnace vessel showed the need to reduce the electrode circle diameter and to increase the arc voltage. The operation with a long arc (at a higher arc voltage) caused a decrease in graphite electrode consumption. Now, it is assumed that the power supply system should operate at such an arc current intensity at which the maximum arc power occurs. However, the operation with higher powers is detrimental to the refractory lining life. Therefore, designs enabling water cooling of the furnace walls and roof to be introduced have appeared. The environmental protection requirements have also forced changes in the furnace design. Dust collection systems were built to enable the off-gases captured in the fourth hole in the roof to be cleaned from dust, and potentially, off-gas heat to be utilized.

In the design and construction of fourth-generation steelmaking EAFs, built in the 1980s [4], special attention was paid to the operator's working conditions and the development of off-gas heat utilization systems as well as automatic control systems and the development of secondary metallurgy. To eliminate the accompanying noise, the furnaces were enclosed in dust- and soundproof chambers. Off-gas exhaust and treatment systems were extended. More attention was paid to charge preparation by more careful scrap sorting and the utilization of off-gas heat for scrap preheating. The heat of the water cooling the walls and roof has also been utilized. Frequently, the arc furnace has only been used for charge meltdown or potentially for the partial oxidation of the metal bath. The other process operations have been transferred to a ladle furnace or other secondary refining facilities. Eccentric bottom tapping is commonly used instead of a traditional system with a pouring spout. It aims at shortening the tapping time, and thereby the reduction of the possibilities for steel stream oxidation with atmospheric air oxygen and facilitating the separation of steel from slag. In order to reduce electrode consumption, their water cooling has been applied. Research on the application of direct current to supply furnaces has also been carried out. The performance results of DC furnaces show a favorable decrease of the harmful impact on the power supply grid as well as a reduction of noise and electrode consumption.

AC EAFs, in particular high power ones, despite many advantages from the perspective of arcing, have some disadvantages that can include a high consumption of graphite electrodes and the refractory lining as well as a detrimental impact on the energy quality in the power supply system. To eliminate these adverse effects, at the turn of the 1980s and 1990s, DC furnaces were introduced into the production practice. There are three fundamental differences in their design compared to the AC furnaces:

- in the power supply system, there is an additional component: AC/DC converter,
- there is a single graphite electrode instead of three,
- a conductive hearth electrode is present in the bottom refractory lining.

Despite the need for the work of an additional component in the power supply system, which is the current converter, and some problems with the hearth electrode operation during the work of DC furnaces, a significantly lower consumption of graphite electrodes and a much lower detrimental effect on the power supply grid are obtained.

DC furnaces have not replaced AC ones, but their number has been growing. For instance, in Japan, there are 22 DC furnaces in operation now, with capacities from 30 to 429 tons [5]. They include one of the biggest arc furnaces worldwide, a DC furnace with a capacity of 420 tons, equipped with a 170-MVA power transformer, at the arc voltage of 600 V and the rated current of 280 kA. The furnace vessel has an internal diameter of 9.7 m and four water-cooled hearth electrodes; it works together with the scrap preheating system of the Consteel type. The common carbon steel production practice includes the meltdown of preheated scrap, with the simultaneous oxidation of the metal bath, followed by tapping and further refining in a ladle furnace. About 33 Nm^3 of oxygen per metric ton of steel produced is consumed in the process, and performance indicators are as follows: tap-to-tap time up to 50 min., electricity consumption 390 kWh/t, electrode consumption 1.2 kg/t, productivity 360 t/h, and meltdown indicator 2.1 t/h and per kWh.

All of the improvements introduced in the design and process practice in EAFs aimed at improving performance parameters, natural environment protection, and the improvement of the operator's occupational safety. The impact of improvements on the crucial performance indicators of an EAF over the past 50 years is shown in Figure 1.4.

These indicators include the electricity consumption, electrode consumption, and the tap-to-tap time. In the 1960s, the following level of these indicators was achieved:

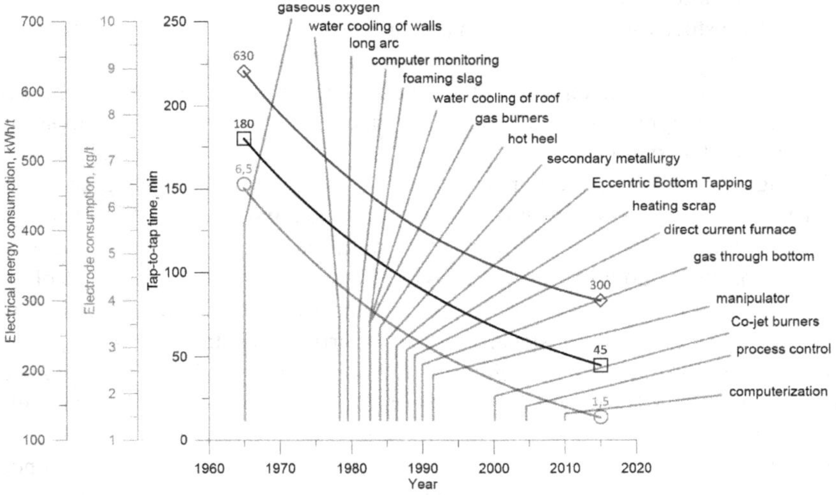

FIGURE 1.4 Impact of technological and technical improvements in the EAF operation on the crucial performance indicators over the past 50 years. (Author's work.)

- electricity consumption of 630 kWh/t,
- graphite electrode consumption of 6.5 kg/t,
- tap-to-tap time up to 180 min.

The first significant technological solution improving these indicators was the application of an oxygen gas blowing for the oxidation of the metal bath. In the 1970s, a practice dividing the melting into two stages started:

- charge meltdown in an EAF to obtain a liquid semi-product called metal bath,
- further refining of the metal bath in a ladle furnace (so-called ladle metallurgy) to obtain steel ready for casting.

These two solutions considerably improved performance.

Subsequent improvements included the introduction of water cooling of the walls and roof, increase of furnace transformer power, introduction of foamy slag practice, eccentric bottom tapping, gas-oxygen burners, etc. Thanks to all of these solutions, modern steelmaking shops can now achieve the following EAF performance indicators:

- electricity consumption of 300 kWh/t,
- graphite electrode consumption of approximately 1.5 kg/t,
- tap-to-tap time approximately 45 min.

1.4 DEVELOPMENT OF ELECTRIC ARC FURNACE PRACTICE

The versatile and fast development of technology in the 19th century required supplies of more steel with high functional properties. First, these needs were satisfied by supplies of steels made in the crucible and acid open-hearth process. However, these processes required applying a metallic charge with low phosphorus and sulfur contents, which significantly influenced the steel manufacturing costs and substantially restricted its output.

During World War I, the demand for quality steel was increasing so fast that the crucible process was not able to satisfy this demand. It caused the transfer of the production of this steel to EAFs that were already in operation. However, the first trials of making quality steels in EAFs did not give positive results. Tool steels made in these furnaces had a very limited hardness range, and structural steels had a low toughness. In addition, structural steels, particularly steels for the armaments industry, showed defects in the form of hydrogen flakes, which in the acid crucible and open-hearth steel occurred very rarely and in small amounts. This effect was incomprehensible and was wrongly related solely to the contents of oxygen, phosphorus, and sulfur in steel. Only O. Thallner observed that the occurrence of flakes was related to the hydrogen content in steel, and its properties also depend on the nitrogen content. The promising application of white and carbidic slag as suggested by F.R. Bichhoffs made conditions for the production in EAFs of steel, which in terms of quality was as good as the crucible steel but had much lower manufacturing costs. Mastering the steelmaking manufacturing process in EAFs, combined with the construction of electricity transfer lines and the accumulation of large amounts of scrap and its favorable prices, has led to the development of steel electrometallurgy.

Taking into account the complete elimination of the open-hearth process, and consequently the necessity of processing the entire surplus scrap in electric furnaces, the share of steel produced in them in the global production will increase up to approximately 35%. It is the top limit of steel production in electric furnaces arising from the scrap balance in the global industrial economy, and the possibilities of its use in oxygen steelmaking processes. In recent years, positive results have been obtained regarding the application of other materials as the charge for EAFs. Practically, materials from direct reduction processes are more often applied, and there are also attempts to use liquid charge (hot metal). This encourages the development of steelmaking processes in electric furnaces. An additional factor determining the development of this method of steel production is the ease of full control of operating parameters. The optimization of manufacturing processes and control of the power supply system parameters as well as the related control of electrode feed is a common practice nowadays.

One of the most important achievements in the area of steelmaking is the development and acceptance of methods of secondary metallurgy. In modern metallurgy, a steelmaking furnace is only used to obtain a liquid metal bath as a semi-product for further refining in a ladle, and in secondary metallurgy facilities. This practice definitely improves the quality of the steel manufactured, increases productivity, and at the same time maximizes cost-efficiency of the production process. The reduction of tapping temperature and making all the refining operations beyond the EAF allows the tap-to-tap time to shorten, which reduces heat losses and the consumption of refractory materials [5].

The basic parameter characterizing the EAF is its size (capacity), or the mass of liquid metal that is accommodated in the vessel during a standard smelting, which can be obtained and transferred for further process operations after having the completed smelting procedure (so-called tapping). Metal is tapped from the furnace into a steel ladle. In terms of size, we can distinguish small capacity furnaces (up to 20 tons of metal), medium capacity furnaces (from 20 to 80 tons), and large capacity furnaces (over 80 tons). The furnace capacity often determines the range of steel grades produced. In small- and medium-sized furnaces, usually medium- and high-alloyed steels are produced due to a better flexibility for small orders. However, large capacity furnaces are rather used for the production of common carbon steels. The annual production capacity of a steelmaking shop equipped with an EAF is from a few dozen metric tons per annum, for small furnaces, up to 1–2 million tons for large furnaces. The biggest arc furnace in operation now has a capacity of 420 tons.

The development of steel production worldwide from the beginning of the 20th century until 2019 is shown in Figure 1.5. The production volume in 1900 was approximately 28,000 metric tons, whereas in 2019, it was 1,817 million tons (according to the Worldsteel Association). In global statistics, the share of EAF steel production in

FIGURE 1.5 Total volume of steel production worldwide from 1900 [6]. (Author's work.)

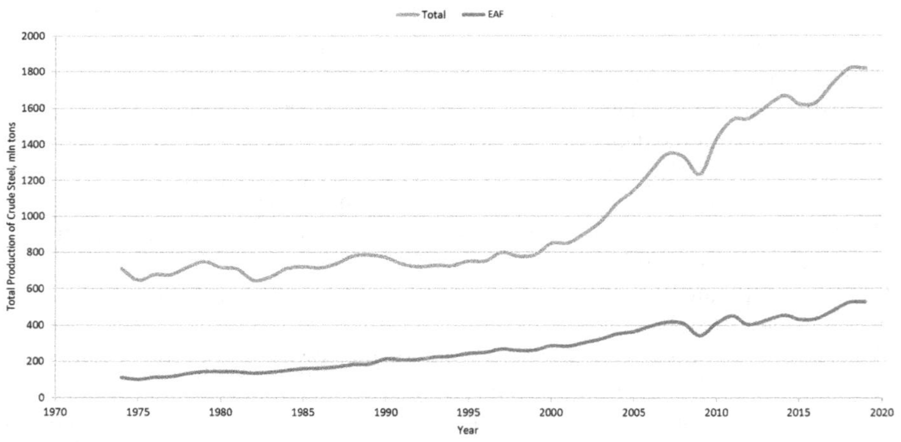

FIGURE 1.6 Steel production in the world: total and in EAFs from 1973 [6]. (Author's work.)

the total output were noted only in the mid-1970s. At the beginning, this share was approximately 16% (Figure 1.6). In terms of quantity, the increase of steel production in EAFs was relatively slow, but it was faster than the increase in the total production. In late 20th century, the share of this production reached over 33%. The growth was faster primarily in those countries where the electric energy was cheaper. However, in recent years, the rapid growth in demand for steel has caused an increase in the BOF production, and the share of EAF production has declined to approximately 27% in 2019.

REFERENCES

1. Ayrton H.: *The Electric Arc*, D. Van Nostrand Co., New York, 1902.
2. Taylor C. R.: *Electric Furnace Steelmaking*, The Iron and Steel Society, New York, 1985.
3. Ameling D.: Steel – Innovative solutions for energy and resource challenges, *9th European Electric Steel Congress*, Krakow, Poland, 2008.
4. Jones J. A. T.: EAF steelmaking – Current state of the art technology and future developments, *18th Steelmaking Conference*, IAS, Rosario, Argentina, 2011.
5. Emi T.: Steelmaking Technology for the Last 100 Years: Toward Highly Efficient Mass Production Systems for High Quality Steels, *ISIJ International*, Vol. 55, 2015, No. 1, pp. 36–66.
6. Steel Statistical Yearbook, World Steel Association. https://www.worldsteel.org/steel-by-topic/statistics/steel-statistical-yearbook.html.

2 Layout of an Electric Furnace Shop

The buildings and bays of an electric furnace shop are designed to ensure that the manufacturing process is properly organized and the assumed performance indicators are obtained, while the transportation and logistics system should be as advantageous as possible, which largely determines the production cost minimization. The arrangement and equipment of an electric arc furnace (EAF) shop depend on the designed manufacturing capacity, type of the charge processed, range of the steel grades manufactured, and the melting and casting practice. The manufacturing capacities of EAF shops range from tens of thousands up to two million metric tons of steel per year. The following materials are used as the iron-bearing charge: steel scrap or the so-called direct reduction iron such as DRI and HBI. The type of charge materials applied is the main factor that determines the layout and design of the charging bay as well as the design of the furnace bay. The range of steels manufactured determines the types of practices applied as well as the related specialist auxiliary equipment. It is also decisive for the arrangement of this equipment in the charging bay, furnace bay, and casting bay. The steel casting method (either continuous or ingot casting) determines the design solutions of the casting bay and its equipment.

When designing an electric furnace shop, its buildings and bays should be situated in the general steel plant layout to ensure that the transport between departments is rational and in line with the principles of logistics. The design solution of the main EAF shop building depends on the furnace (or furnaces) location, arrangement of furnace's individual components, and the number and types of other devices. Now, the so-called mini-mills are the most common. They comprise a single manufacturing line, including a single EAF as the basic production unit, secondary steelmaking units and casting units, and auxiliary equipment, which supports the operation of those units. Steel mills that include more than one furnace are less common. One of the most important design assumptions of a steelmaking shop building is the method of furnace foundation. Two solutions can be applied: either the foundation with the operating door at the zero level of the bay or the foundation on a structure a few meters above the zero level, on the so-called main platform. In the first case, the building is lower and the investment costs are lower, but solutions related to the arrangement of the electric equipment or metal tapping, etc., are more difficult. The other advantageous solution is more common. Here, the furnace foundation is located at the zero level, and the vessel is a few meters above. In this case, a steel structure called the main platform is built a bit under the operating door level. The following are placed on the platform: operator's cabin, manipulator, and other auxiliaries. It is the main operating level for the furnace operation.

DOI: 10.1201/9781003130949-2

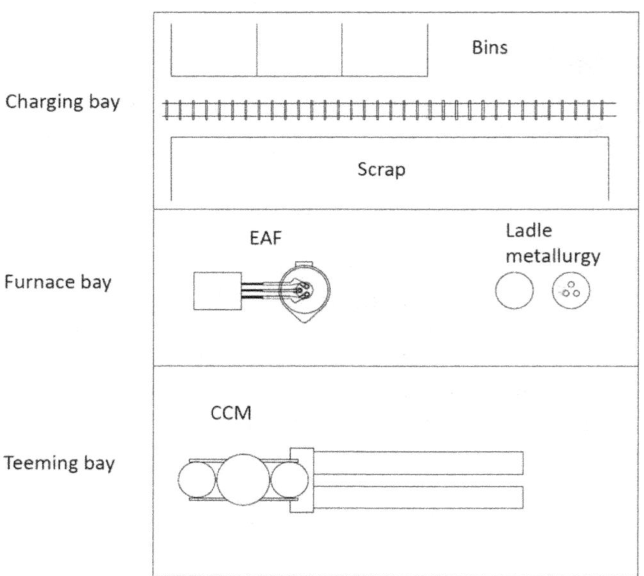

FIGURE 2.1 General layout of an EAF shop building with the bay arrangement. (Author's work.)

The main steel shop building usually consists of three parallel bays: charging bay, furnace bay, and casting bay. The general layout of a steelmaking shop is shown in Figure 2.1.

The size of individual bays and the communication system between them depend on the individual needs and conditions, and they are primarily adjusted to the production capacities and the range of the steels manufactured.

As indicated by its name, in the charging bay, charge materials are received, stored, and prepared to be loaded into the furnace. The bay is usually rectangular, it is 20 m–30 m wide, and its area is adjusted to the production capacity. It is equipped with a few overhead cranes traveling along the longer axis of the bay, railway tracks, and a road for rail and truck transport as well as bins for scrap and other metallic materials, slag-forming materials, and carburizing materials. The number of bins is primarily related to the number of scrap types processed, and their capacity should be sufficient to store the charge materials necessary for 10–30 days of continuous operation. Charge materials can be delivered by rail or road situated along the longer axis of the bay. Charge materials are unloaded from railway cars or trucks and subsequently loaded into charging baskets with cranes fitted with lifting magnets or clamshell grabs. The general layout of a charging bay is shown in Figure 2.2.

Storage bunkers and bins for individual types of materials are arranged along the longer sides of the bay, on both sides of the railway tracks. Segregated types of scrap are placed in bunkers, whereas other materials – slag-forming, carburizing, ferroalloys, etc. – are placed in bins. These materials are usually delivered in special containers and are unloaded with overhead traveling cranes fitted with cross-bars adapted to handle these containers. In modern design solutions, bins for slag-forming

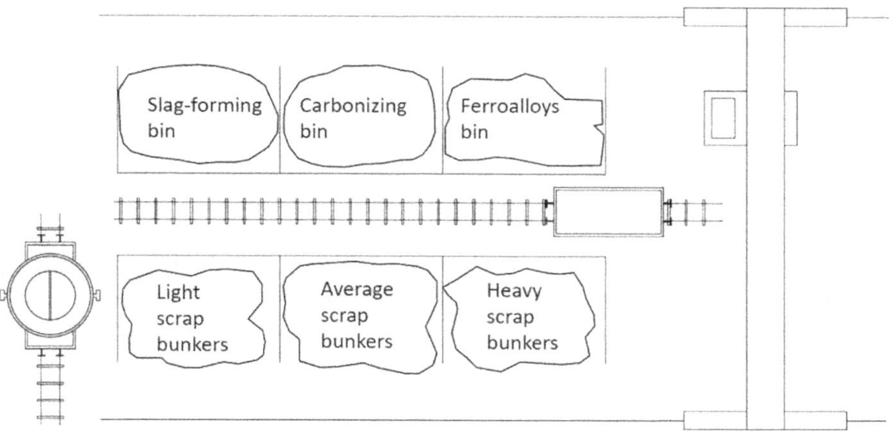

FIGURE 2.2 General layout of a charging bay. (Author's work.)

and carburizing materials as well as ferroalloys are arranged in series in the furnace bay above the main platform level. These bins are loaded from the top. Their lower part has a shape of a hopper ended with a feeder that allows for smooth control of the amount of material discharged. A belt conveyor on which a prepared portion of relevant material is discharged according to the demand is placed underneath the bins. The belt conveyor transports the prepared materials either to the charging bay to load them into a charging basket or to the furnace bay to load them directly into the furnace. Bins with basic ferroalloys and graphite electrodes are situated at the other end of the charging bay. Crane tracks are located at a height of over 10 m above this level along the longer walls of the charging bay. The crane bridge moving on the rails is equipped with a cart containing a lifting mechanism. The grabber of the lifting mechanism enables to move materials vertically and horizontally, in the space limited by the length of the track (usually the same as the bay length), the width of the bridge (usually the same as the bay width), and the height of lifting and lowering (usually the same as the bay height). Usually, a few independent cranes travel on common rails in order to ensure the efficient unloading of the materials handled and loading them into a charging basket.

Most often, in the half of the length of the bay, in the "scrap" part, there is a stand, on which a mobile cart with a basket is placed for loading a metallic charge into it. The baskets of clamshell or the orange peel type are placed on a special car located on the tracks. These tracks are located in the direction perpendicular to the main axis of the charge bay, and they enable the basket to be transferred to the furnace bay. The cars are usually equipped with electronic scales that enable ongoing weighing of the material to be loaded. In addition, modern visualization techniques are introduced for the loaded scrap in order to optimize the scrap distribution in the basket for minimizing the electricity consumption during meltdown [1]. The scrap is loaded into the charging basket using cranes with lifting magnets, which transfer the scrap from the bins into the basket. Slag-forming and carburizing materials are transported from the relevant bins containing these materials with belt conveyors, from which they are fed into the basket with a chute.

The furnace bay has a design similar to the charging bay, and it is situated in parallel. Overhead cranes are located along its main axis, and they also travel along the longer axis of the bay. The cranes ensure the loading of baskets into the furnace, transport of ladles during tapping, and casting of metal as well as transport of various other materials. At least one crane must have a sufficient capacity to lift both a basket full of charge and a ladle full of tapped steel, including the weight of the ladle. This bay houses an EAF or furnaces, if more than one, secondary metallurgy equipment (e.g., a ladle furnace), and sometimes a vacuum refining stand or other auxiliary equipment or stations for the manufacturing process operation. Both the furnace and all of the other equipment are arranged in the same line along the longer axis, usually next to the wall adjacent to the charging bay. The general layout of a furnace bay in the horizontal and vertical plane is shown in Figures 2.3 and 2.4, respectively.

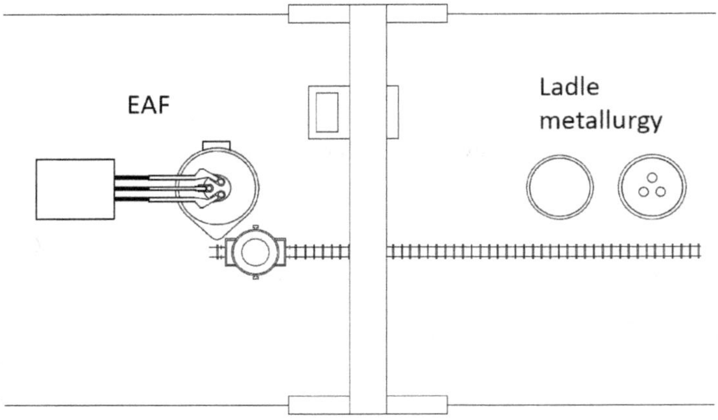

FIGURE 2.3 General layout of a furnace bay, plan view. (Author's work.)

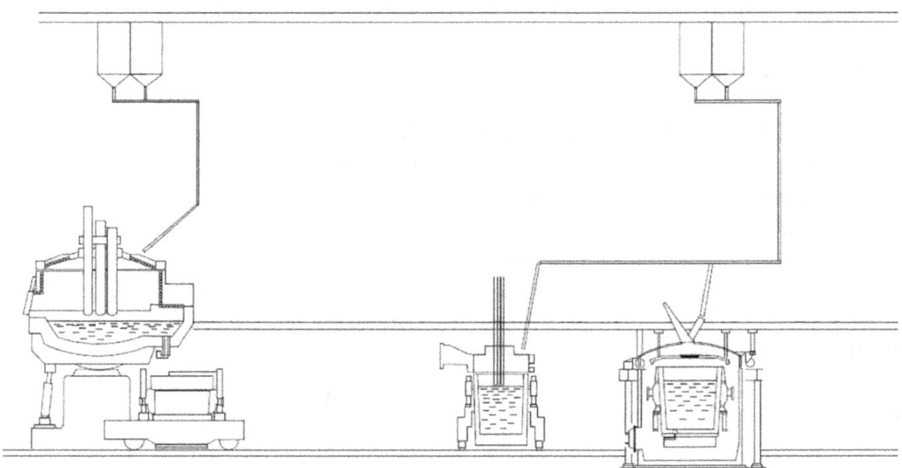

FIGURE 2.4 General layout of a furnace bay, side view. (Author's work.)

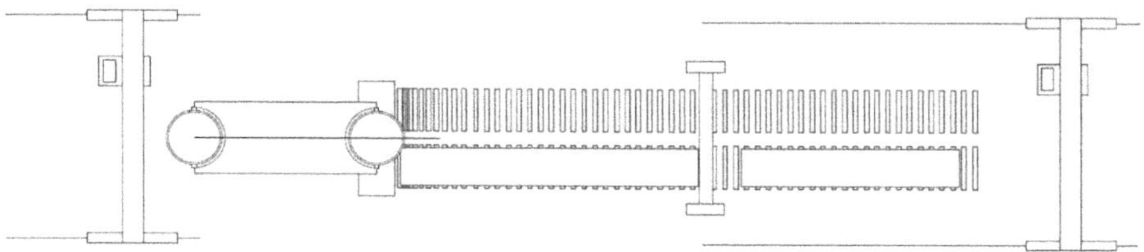

FIGURE 2.5 General layout of a casting bay. (Author's Work.)

The furnace bay usually has two parallel parts along the longer axis, with various operation levels. At the zero level, in the part closer to the casting bay, there is a system of tracks for transporting steel and slag ladles. The tracks must enable a steel ladle to be transported longitudinally, between the arc furnace and the ladle furnace, and potentially between other workstations. Transverse tracks enable the steel ladle to be transferred to the casting bay. The tracks for slag ladle transport are usually located perpendicular to the longitudinal axis of the bay. At the main platform level, only the charging basket, electrodes, and auxiliary materials are handled with the appropriate cranes traveling longitudinally to the bay. In this part, the basic technical and technological operation of the arc furnace and secondary metallurgy equipment is performed. At this level, there is a transformer and the power supply system, operator's control cabin, electrode assembly station, vessel and roof repair stands, and other auxiliary stations.

The design and equipment of the casting bay are related to the type and size of the continuous casting machine. This bay adjoins the furnace bay. The general layout of a casting bay is presented in Figure 2.5.

In some cases, casting of steel is done into ingot molds. Then, overhead cranes are placed in the casting bay, traveling along the longer axis of the bay, thereby ensuring the relocation of the ladle during steel casting. Also, along the longer side, there are ingot mold stations to which steel is poured. The stations may be located in casting pits; then, the upper ingot mold level is at the zero level, where the casting process is operated. A special service level at a height of the upper mold level is another solution. Then, the molds are placed at the zero level and they are handled from the service level. In addition, ladle repair stations are located in the casting bay at a face wall. The refractory lining is laid at these stations. Then, it is dried and heated; also the old lining is demolished. Spare molds are placed next to the other face wall, and there is a station for their repair as well as potentially a place to store the ingots removed from the molds. Often, to save energy, ingots with molds are transported immediately after solidifying (in the so-called hot condition) to plastic working departments: forging or pressing.

Apart from the main building, the steelmaking department also includes office buildings, staff facilities, and workshops for mechanical and electrical repairs for the purposes of the so-called maintenance.

REFERENCE

1. Baumert J. C., Picco M, Weiler, C, Wauters M, Albart P, Nyssen P.: Automated Assessment of Scrap Quality before Loading into EAF, *Archives of Metallurgy and Materials*, Vol. 53, 2008, No. 2, pp. 345–351.

3 Construction of Electric Arc Furnaces

Electric arc furnaces (EAFs) consist of three distinct groups of equipment: a furnace vessel with a roof, mechanical equipment, and electric equipment. The metallurgical steelmaking process takes place in the vessel; therefore, it is its main structural component. The vessel comprises a steel shell, lined on the inside with refractories, constituting a reaction chamber closed in with a roof. The roof has holes for electrodes and gas evacuation. The main mechanical equipment of an EAF includes various mechanisms for furnace tilting, lifting, and lowering the roof, electrodes, and the main door bolt as well as the off-gas discharge and dust removal systems. The electric equipment includes a furnace substation along with a transformer, secondary circuit, systems of automatic control of electrode positions and power input, and the control cabin accessories. The EAF design is shown in an axonometric projection in Figure 3.1.

On the right-hand side of the figure, you can see the furnace vessel with its hearth (also called bottom). During melting, the liquid metal is in this space. The other part of the furnace is above – the walls made of water-cooled elements. The roof is placed on the furnace vessel. It is also made of water-cooled elements, and in its central part, it has a separate centerpiece made of refractory materials, with three holes for inserting graphite electrodes. The furnace is mounted on a concrete foundation, where it is

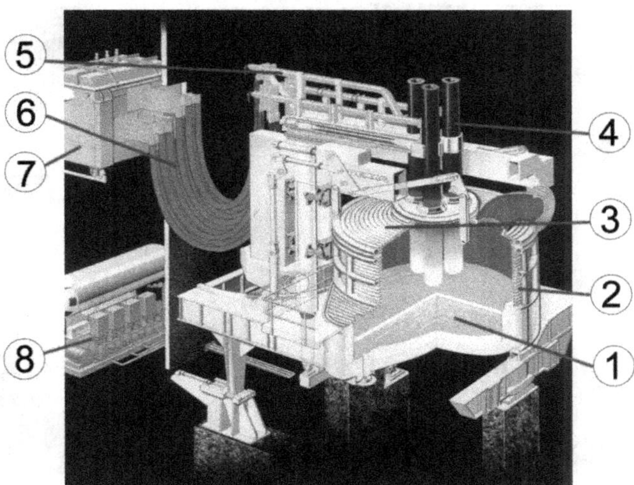

FIGURE 3.1 General view of an electric arc furnace in an axonometric projection [1]: 1 – bottom, 2 – water-cooled walls, 3 – roof, 4 – graphite electrodes, 5 – tubular buses, 6 – flexible cables, 7 – transformer, 8 – hydraulic oil tank. (Advertising materials (corporate brochure) of the SMS group GmbH (formerly SMS Siemag AG) – with permission.)

DOI: 10.1201/9781003130949-3

placed by means of a steel structure shaped as two rockers. An operating platform steel structure is secured to the rockers, along with a mast being a load-carrying structure for the roof and the electrodes. All of these components form a whole in the design sense, and the rockers enable it to be tilted with hydraulic cylinders placed underneath.

The graphite electrodes are inserted into the furnace from the top, through holes in the roof. The electrodes are the final element of the power supply system. During the furnace operation, electric arcs burn at the electrode tips, providing the thermal energy necessary for the execution of the steelmaking process. The electrodes are held with load-carrying arms secured to a vertical mast. The mast, along with the arms, can shift in the vertical plane and partially swivel in the horizontal plane, which enables the roof to be swung together with the electrodes during charging, and the electrodes to be shifted during the charge melting.

Conductors supplying power to the electrodes (so-called tube conductors) are fixed to the top part of the electrode support arms. The so-called flexible cables, which are connected to the tube conductors, comprise a "U"-shaped body in order to make a flexible connection between the movable electrodes and the stationary furnace transformer. The transformer seen in the top left corner of the picture is situated near the furnace, usually in a separate room. A hydraulic oil supply station can be seen in the bottom left part of the picture. It supplies oil to hydraulic cylinders driving all the moving components of the furnace.

Pictures of a furnace in two views – "side" and "top" – are shown in Figures 3.2 and 3.3. In the central part of both pictures, you can see the furnace with its roof,

FIGURE 3.2 General view of an electric arc furnace from the main door (also called slag door) side [1]. (Advertising materials (corporate brochure) of the SMS group GmbH (formerly SMS Siemag AG) – with permission.)

FIGURE 3.3 General view of an electric arc furnace from the top [1]. (Advertising materials (corporate brochure) of the SMS group GmbH (formerly SMS Siemag AG) – with permission.)

whereas on the right-hand side, there is the mast with the roof and the electrode carrying arms. On the left-hand side, you can see the system of gas and dust evacuation from the reaction chamber, for the purposes of natural environment protection.

Figure 3.4 shows a cross-sectional diagram of an EAF construction, including pictures of the selected structural components.

3.1 SHAPE AND DIMENSIONS OF THE REACTION CHAMBER

The EAF reaction chamber usually has a circular horizontal cross section. Its dimensions are primarily determined by the furnace capacity, its design and equipment, and the type of manufacturing process and the range of steels manufactured.

The hearth (bottom) has a conical-spherical shape, and its volume is calculated on the basis of the mass of steel that is to be produced in a single heat, and the related mass of slag. Figure 3.5 shows a diagram of a cross section of the furnace vessel showing its shape and basic dimensions.

The hearth volume can be calculated with a simplified formula [2]:

$$V = G \cdot \left[0.145 + a \cdot (0.3 - 0.35)\right] \tag{3.1}$$

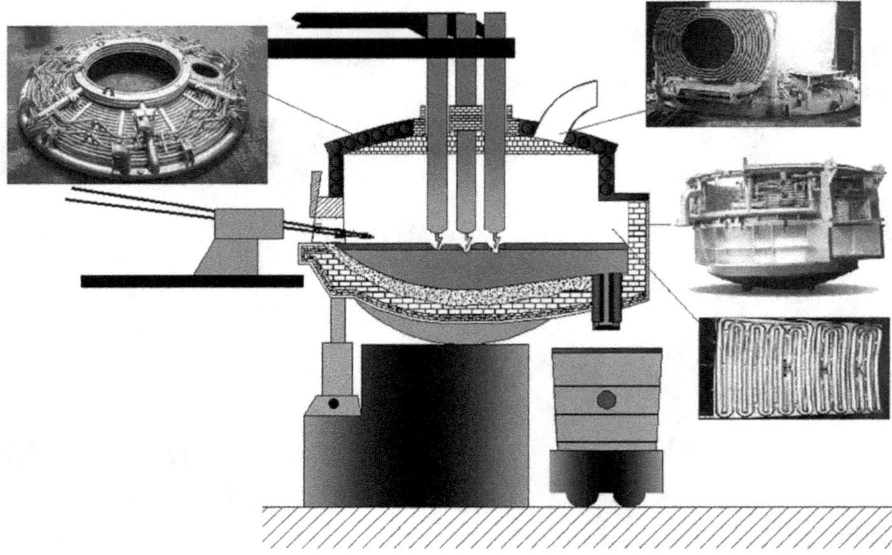

FIGURE 3.4 Cross-sectional diagram of an electric arc furnace construction, including pictures of the selected structural components. (Cross section – author's work.)

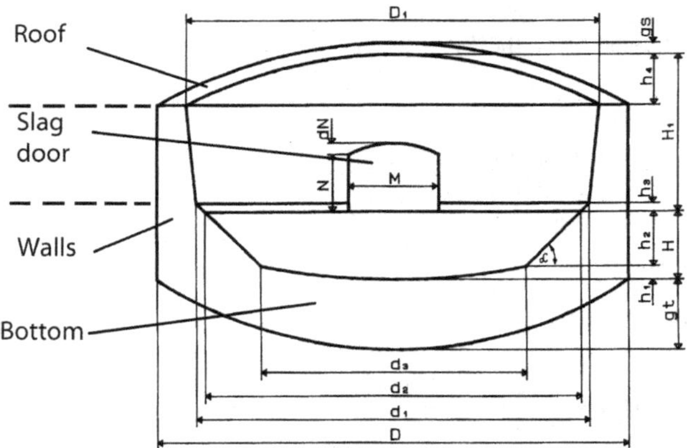

FIGURE 3.5 Diagram of a cross section of an electric arc furnace vessel. (Author's work.)

where:

V – calculated hearth volume, m^3,

G – furnace capacity (mass of liquid steel in a single heat), t,

a – share of the slag mass in relation to the metal mass, it is assumed $a = 0.05–0.10$.

In the above formula, number 0.145 is the volume of 1 t of liquid steel (average) expressed in m^3/t, whereas the number from the range of 0.3–0.35 means the volume of 1 t of liquid slag, also expressed in m^3/t.

When determining the basic dimensions of the furnace hearth, it is assumed that the diameter to depth ratio is from 4.0 to 7.0. Values below this ratio refer to smaller sized furnaces, which stems from the need to obtain a sufficient metal bath depth to protect the bottom shell against overheating by electric arcs. For larger sized furnaces, higher values of this ratio are applied, but it is assumed that the maximum depth of the metal bath should not exceed the limit of 1.2 m.

The angle α between the conical part of the hearth and the horizontal plane should be from 30° to 45°. In small- and medium-sized furnaces, it is usually 45°, and for larger sized ones, it is 30°. This shape of the hearth ensures that the metal and slag completely flows down during tapping, for the furnace tilt angle of 40°–42°. The height of the conical part of the hearth is usually 4/5 of its depth, and for the spherical part, it is 1/5 of the depth. The hearth volume expressed with its geometrical dimensions is determined by the following formula [2]:

$$V = \frac{1}{6}\pi \cdot h_1 \left(3\frac{d_3^2}{4} + 3h_1^2 - 2h_1 \right) + \frac{h_2}{12}\left(d_2^2 + d_2 \cdot d_3 + d_3^2 \right) \tag{3.2}$$

where:

d_2 – diameter of the furnace reaction chamber at the level of the main door sill, m,
d_3 – smaller diameter of the conical part of the hearth, m,
h_1 – height of the spherical part of the hearth, m,
h_2 – height of the conical part of the hearth, m.

Using equation (3.2), the angle α, and the relationship between the melt depth and the height of the conical and spherical part of the hearth, we can compute its necessary volume for a given furnace capacity, and subsequently – using these values – the other dimensions.

From the process engineering perspective, the height, or the distance from the main door sill to the roof base, is an important dimension of the reaction chamber. For large reaction chamber heights, the roof is more distant from the electric arcs and the metal bath, which increases its life. An excessive height of the chamber increases heat losses and contributes to an increase in the unit electricity consumption. The literature provides the following empirical formula for determining the reaction chamber height, depending on the furnace capacity [2]:

$$H_1 = 0.75 \cdot G^{0.25} \tag{3.3}$$

It is shown in formula (3.3) that for small- and medium-sized furnaces, as their capacity increases, the reaction chamber height grows quickly. That is related to the impact of electric arc radiation on the roof. However, as the distance between the roof and the electric arcs increases, their mutual influence gradually decreases. Therefore, an increase in the larger sized furnace capacity has a relatively low impact on the reaction chamber height. An increase of this height is determined primarily by the power of electric arcs.

More specific details about designing the EAF reaction chamber can be found in the references [2].

3.2 FURNACE SHELL

The furnace shell is a load-carrying component for the bottom lining and potentially for the wall refractory lining. In older design furnaces, it is a single component comprising a cylindrical or cylindrically conical (taper downward) upper part and a spherical (dome-shaped) bottom. The furnace hearth and banks are placed on the shell bottom. They are made of refractory materials, and together they constitute a vessel for liquid metal and slag.

The whole shell structure carries static loads from its mass, lining mass, mass of metal and slag, roof, and dynamic loads occurring during charging. The structure is made of welded steel plates, with a thickness of 20–40 mm (depending on the furnace capacity). Due to higher mechanical loads, the shell bottom is thicker by 2–5 mm. The shell structure is reinforced with vertical and horizontal ribs welded to its external part. Not only does it transmit the stress resulting from mechanical static and dynamic loads, but it also transmits the tensile stress from the furnace lining thermal expansion and the thermal stress caused by the heterogeneity of the temperature field.

There are two holes cut out in the upper part of the furnace shell: for the main door and for the pouring spout. The reaction chamber usually has a circular cross section. The dimensions of the reaction chamber are primarily determined by the furnace size, process type, steel grades manufactured, and the furnace design and equipment. The main door is used for performing some process actions by operators, including measuring the temperature, taking samples, inserting oxygen and coal lances, and a gas burner. The main door hole is braced with a special flange cooled with water. The pouring spout is made of steel plate, and it has the shape of a tapered trough, with refractory lining inside. It is used to guide the liquid metal flowing out from the furnace during tapping to the ladle.

In the majority of modern furnaces, the bottom part of the shell is a separate component, while the walls consist of water-cooled segments with no shell. Both parts of the furnace are connected with a specially profiled flange. During a repair, each of the parts is replaced individually.

Only older design furnaces have a hole for a pouring spout. Modern EAFs have a taphole in the bottom, the so-called eccentric bottom tapping (EBT). In this case, the shell has the shape of a dome and a "protruding" balcony part. Figure 3.6 shows a diagram of a furnace bottom part – EBT type, in a top view, whereas Figure 3.7 shows its side view.

FIGURE 3.6 Top view of furnace bottom – EBT type [1]. (Advertising materials (corporate brochure) of the SMS group GmbH. (Formerly SMS Siemag AG) – with permission.)

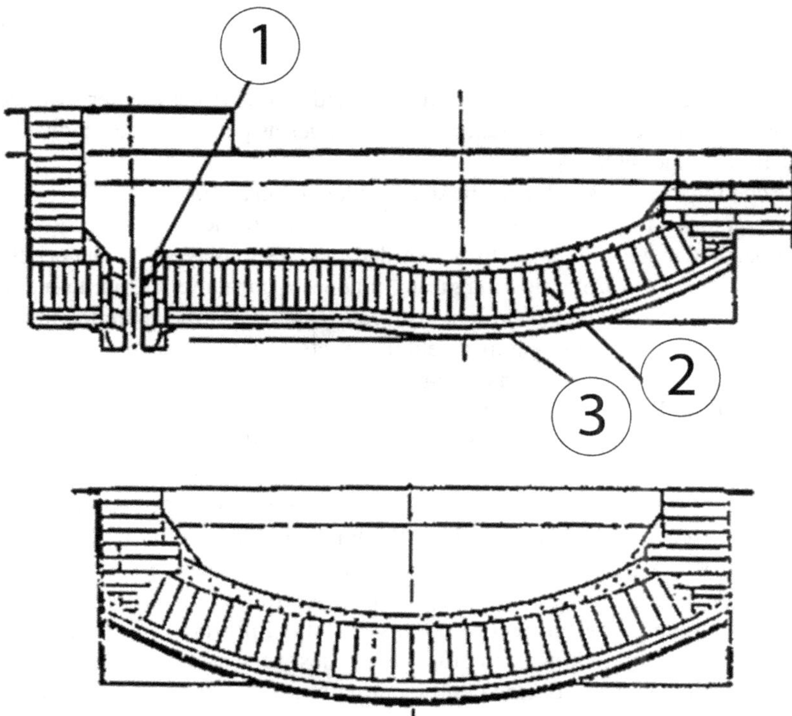

FIGURE 3.7 Side view of furnace bottom – EBT type: 1 – taphole, 2 – bottom refractory lining, 3 – bottom shell. (Author's work.)

3.3 FURNACE REFRACTORY LINING

During steel making in EAFs, the refractory lining is exposed to the corrosive impact of the melt and slag, metal oxide vapors, intensive electric arc radiation, mechanical impact, etc.. Electric arcs are point-type radiation sources, with varied distance to the individual fragments of the refractory lining, which considerably influences the heterogeneity of the lining temperature field. Over the years, there has been a significant progress in reducing the arc power asymmetry and power control. Nevertheless, these factors still considerably affect the heterogeneity of the refractory lining temperature field. The refractory lining is also exposed to frequent, high-temperature variations during fettling, charging, and tapping.

Therefore, the operating conditions of the EAF refractory lining should be considered very harsh. At the same time, the performance of EAFs, the quality of the steels manufactured, and the steel manufacturing costs depend on the lining life. Each EAF has its own characteristics and the expected range of steels manufactured, and they are generally called furnace operating conditions. The furnace refractory lining should be properly adapted to these operating conditions, on the basis of previous operating experience. Minimizing the share of refractory materials in the steel manufacturing costs while ensuring a high furnace performance and a good steel quality is a decisive factor influencing the choice of a specific type of refractories.

Refractories applied for the furnace vessel lining, due to their operating conditions, should feature as high chemical resistance to the impact of liquid metal and slag as possible, as high mechanical strength and refractoriness under load as possible, as low volume changes at temperature changes as possible, and as low thermal conductivity as possible [3].

The chemical processes during the steel production in an EAF are basic. Therefore, basic refractories are also common for the lining in contact with the liquid metal, and in particular with the slag. In recent decades, various types of materials have been applied such as dolomite, magnesia, magnesia-chromite, chromite-magnesia, and spinel materials. Due to the quality requirements for the manufactured steel grades changing over time, environmental aspects, business factors or, finally, the development of refractory materials magnesia-carbon materials have been prevailing in working layer linings of EAFs for about 20 years [4].

Incorporating various forms of carbon to the products made on the basis of magnesia raw materials allowed magnesia materials to obtain new functional features. Primarily, their corrosion resistance has been improved. Carbon is added to magnesia-carbon materials primarily in the form of flake graphites. Thanks to the application of flake graphites with various purities and particle sizes, it was possible to limit the rate of working lining wear. Due to the poor wettability by liquid oxides, graphite provides protection against the slag infiltration into the refractory material. This feature, along with other properties, such as a low linear expansion coefficient, a low sliding friction coefficient with grains of sintered or fused MgO, easy deformation at strain, and anisotropic and flaky structure, gives MgO-C products features that are desired when applying them as a refractory lining for an EAF as well as for some other devices used for the production of iron and steel. The graphite structure and, in particular, its lack of solubility and weak wettability by oxide liquids have caused a different nature of material wear than that observed so far. Burned substances had a disadvantage, which was a low resistance to their infiltration by liquid oxide substances and the formation of zoned structure with varied properties, which resulted in structural spalling. Adding carbon allowed these defects to be eliminated, and it restricted the wear of basic materials in contact with basic slags to local areas, not exceeding a few millimeters. The graphite flaky structure and its orientation in the product microstructure also allow for materials with a very low open porosity, often under 2%, to be obtained. This gives additional properties improving the corrosion resistance. A very low resistance of graphite and other carbon-bearing components to oxidation remains their main disadvantage.

The magnesia-carbon material practice did not change since it had been introduced until the end of the 20th century. Refractory materials were made of clinkers or magnesia-fused aggregates, flake graphites, and other carbon-bearing substances, usually with metallic or nonmetallic anti-oxidizers. Phenol-formaldehyde resins were used as binders. The beginning of the 21st century is an important stage in the development of the MgO-C refractory materials. Liquid carbon pitches started being used as binders. Therefore, the technology of molding mixes has changed, as molding products with carbon pitch binders require temperatures between 363 K (90°C) and 383 K (110°C). At present, MgO-C materials with a pitch binder constitute 80% of all molded basic carbon materials. Regardless of the binder applied in magnesia-carbon products, these products are not burned, but they are typically subjected to drying operations or other heat treatment at a temperature of a few hundred Celsius degrees.

Contrary to the working layers of the furnace refractory lining, magnesia products prevail in protective layers. Magnesia clinker is an input material for manufacturing magnesia materials. The quality of the clinker and clinker materials depends not only on the high MgO content but also on the low content of Fe_2O_3 as possible and the sum and the ratio of CaO and SiO_2 contents. In addition, the method of clinker burning highly affects the quality of magnesia materials. Although magnesia products are very important from the perspective of the furnace safety and, therefore, the safety of a steelmaking shop, they usually have not much contact with liquid metal or slag. Therefore, they are considered optimal in terms of value for money.

Insulating materials applied in EAFs are usually made of pressed ceramic plates/blocks or ceramic/mineral fiber mats on the basis of alumina and silica.

Examples of basic materials most frequently used for the refractory lining of EAFs are summarized in Table 3.1.

3.3.1 Bottom Refractory Lining

The refractory lining of the bottom should be made of such refractories that, similar to the previously described materials of furnace lining, provide as low heat losses as possible and minimize the hazard of metal "escape" from the furnace. In addition, it should create stable thermal conditions for the course of the steelmaking process,

TABLE 3.1
Summary of the Basic Materials Most Frequently Used for the Refractory Lining of Electric Arc Furnaces (Author's Work)

	Content (%) (Most Frequently)					
Type of Material	**MgO**	**CaO[b]**	**SiO_2[b]**	**Al_2O_3**	**Fe_2O_3**	**Cr_2O_3**
Magnesia (protective layers)	92–96	0.5–2.0	0.3–1.5	0.1–1.5	0.5–2.5	–
Magnesia-carbon[a] (slag zone working layers)	97–98	<1.5	<0.5	<0.2	<0.5	–
Magnesia-carbon[b] (metal zone working layers)	95–97	<2.0	<0.8	<0.4	<0.7	–
Magnesa-chromite with direct bond	70–75	<2.0	<2.0	<10.0	7–14	9–13

[a] The chemical composition is provided for the so-called magnesia part.
[b] For magnesia-carbon products, the desired CaO to SiO_2 ratio should be definitely less than 2.

and a fairly uniform temperature field for the metal bath. In order to meet the above conditions, the bottom refractory lining consists of three layers:

- insulating,
- protective, and
- working,

and each of them plays a specific role.

The insulating layer provides for the basic thermal insulation of the shell so that its outer temperature does not exceed 423 K (150°C). It is placed directly on the furnace shell, and its thickness is usually between 10 and 30 mm.

The protective layer, which is made of a few refractory brick layers (at present usually magnesia bricks), is laid on the insulating layer. The thickness of this layer is usually from 250 to 300 mm. The main purpose of the protective layer is to protect against metal "escape" through the hearth in case of the working layer damage. Therefore, it must consist of at least two rows of bricks laid dry, with empty spaces between them filled with magnesite powder. For the same reasons, bricks in individual rows have misaligned joints. In the upper part, the protective layer merges with the so-called banks, which are made of offset laid magnesia bricks, and next, they merge with the furnace walls. Magnesia bricks have a relatively high heat conductivity; therefore, heat losses through the hearth mainly depend on the quality and primarily on a low thermal conductivity of the insulating layer.

Another hearth layer, which is in direct contact with the liquid metal, is the working layer. It is made from rammed magnesia or magnesia-dolomite materials with a specially selected grain size. After ramming, followed by drying and burning, a monolithic working layer forms with a thickness of usually 400–600 mm. As it is in direct contact with the liquid metal bath, the metallurgical cleanness of the steel produced and safety depend on the quality of this layer. The overall thickness of the hearth refractory lining, depending on the furnace size, is usually from 500 to 1,000 mm. The furnace hearth works in stable thermal conditions, and it is mainly exposed to the erosive impact of the metal bath. The erosive and corrosive impact of slag on the hearth is very limited. The life of a correctly made and operated hearth can last up to a few thousand heats.

In smaller, older design furnaces, a taphole is made in the refractory lining at the level where the banks merge with the wall, opposite to the main door. Tapholes are subject to damage by the steel flowing out during tapping. The life of a taphole in structures of this type can be established at a level of a few decades of heats, bearing in mind that they can substantially differ depending on a steelmaking shop. In the meantime, a running repair is performed as needed. The wear of the taphole is caused by chemical corrosion and erosion by washing. As the taphole is being used, its surface becomes less smooth and the clearance increases or decreases, and so the stream of the outflowing metal has more and more swirls and the tapping time shortens or increases. These effects affect the steel quality and, therefore, the taphole needs to be repaired or replaced [4,5].

In modern design furnaces, a bottom taphole is applied, which is placed directly in the hearth, in the so-called balcony part. This hole consists of two components: made of outer bricks in the shape of a rectangular prism with an internal round hole and inner bricks with the shape of a bushing. In both cases, the bricks are made from very high quality magnesia-carbon materials. The diameter of the taphole for liquid metal is usually from 110 mm to 180 mm, and the brick height is correlated with the hearth thickness. This location of the taphole ensures better conditions for the steel flowing out from the furnace to a ladle. The taphole in the hearth requires the furnace tilt angle between 10° and 15°, whereas for the pouring spout, this angle must be from 40° to 45°. The life of tapholes like these often reaches 150–250 heats.

Arrangement diagrams of the individual bottom layers on the furnace shell, in cases of a furnace with a pouring spout and a furnace with a bottom taphole, are presented in Figures 3.8 and 3.9. One of the potential design solutions of the taphole in the bottom is shown in Figure 3.10.

The furnace bottom may also contain the so-called porous plugs, supplying inert gases to support the stirring of the metal bath. There are solutions with one or more plugs deployed in various places of the bottom (Figure 3.11). Right deployment of the plugs should create as advantageous metal bath stirring conditions as possible. The plugs are made in the form of a cuboidal block from magnesia-carbon refractory materials, with thin pipes for gas flow placed inside in parallel to the longer side. The plug height must be adjusted to the furnace bottom height. Figure 3.12 shows the diagram of the construction and method of securing porous plugs in the bottom.

3.3.2 Wall Refractory Lining

The type of materials applied depends on the local operating conditions, including the range of manufactured steels and melting practices applied. At present, magnesia-carbon materials are almost the exclusively applied type of refractories for the EAF working layers, although in the walls above the slag zone, magnesia-chromite materials can also be encountered. Frequently, in the vessel, the so-called zoned installation is applied, and materials with various quality and strength are used, which stems from different chemical and thermal loads in the individual areas of the walls. Therefore, despite different loads, the wear is as uniform as possible,

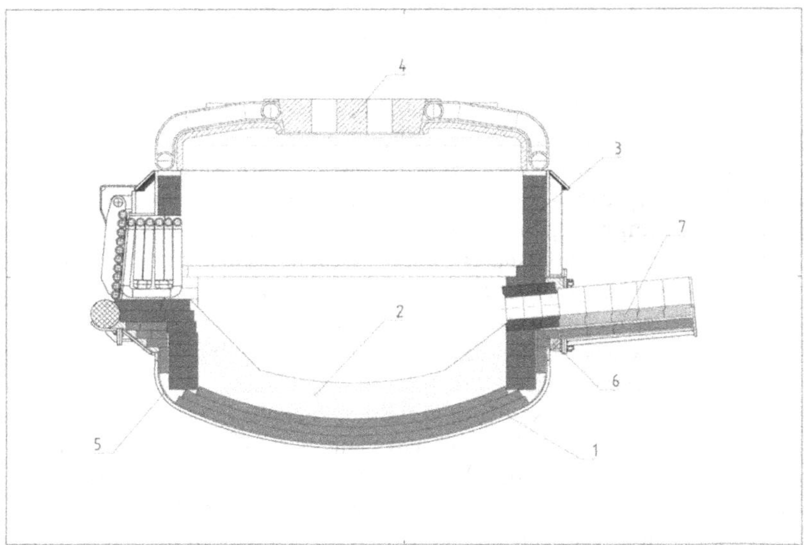

FIGURE 3.8 Example of the arrangement diagram of refractory materials for a furnace with a pouring spout [6]: 1 – protective layer, 2 – bottom working layer made of rammed materials, 3 – furnace wall working layers made of refractory bricks, 4 – refractory concrete roof, 5 – insulating layer, 6 – taphole bricks, 7 – pouring spout. (Advertising materials (corporate brochure) of the Zaklady Magnezytowe "Ropczyce" – with permission.)

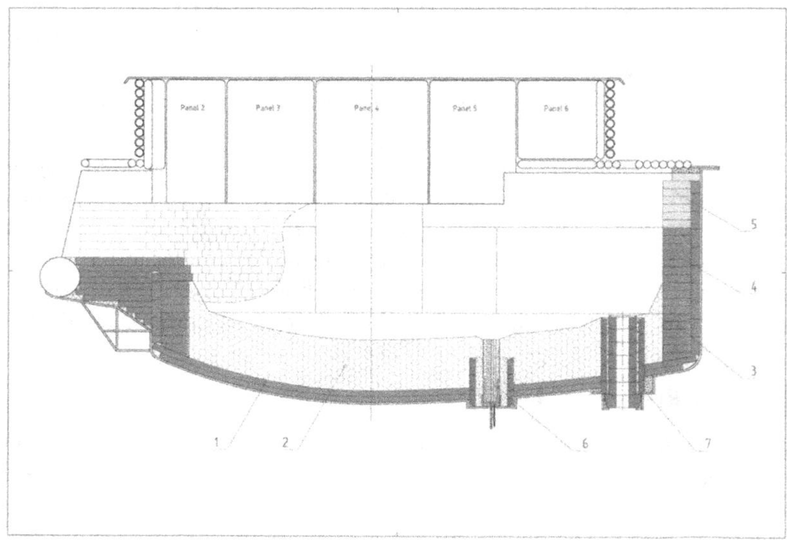

FIGURE 3.9 Example of the arrangement diagram of refractory materials, for a furnace with a taphole in the bottom [6]: 1 – protective layer, 2 – bottom working layer made of rammed materials, 3 – furnace wall working layers in the metal zone, 4 – furnace wall working layers in the slag zone, 5 – wall working layers above the slag zone, 6 – porous plug, 7 – taphole. (Advertising materials (corporate brochure) of the Zaklady Magnezytowe "Ropczyce" – with permission.)

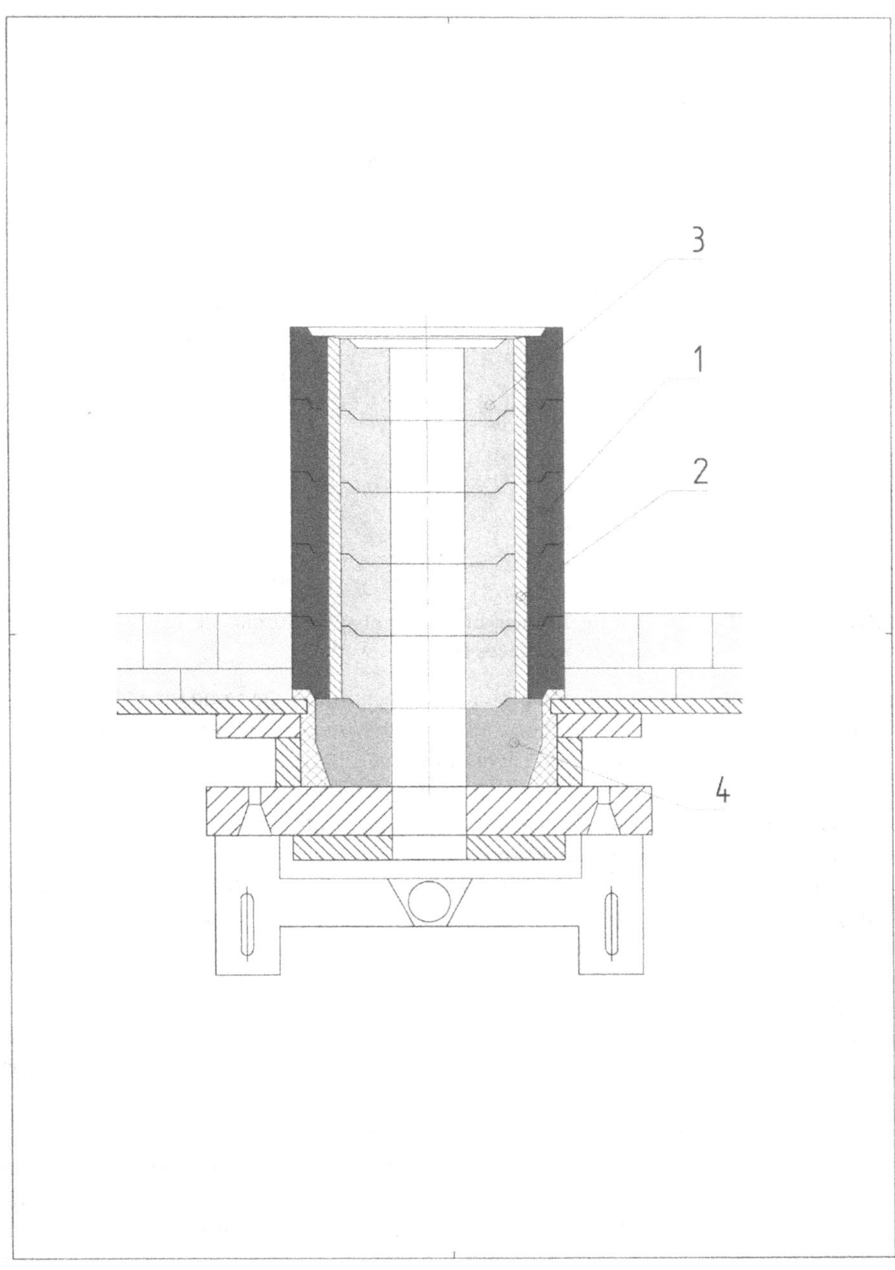

FIGURE 3.10 Design diagram of the taphole in the bottom [6]: 1 – refractory protecting bricks, 2 – gap made of refractory mass, additionally compacted by ramming, 3 – upper bricks of the taphole, 4 – lower bricks of the taphole. (Advertising materials (corporate brochure) of the Zaklady Magnezytowe "Ropczyce" – with permission.)

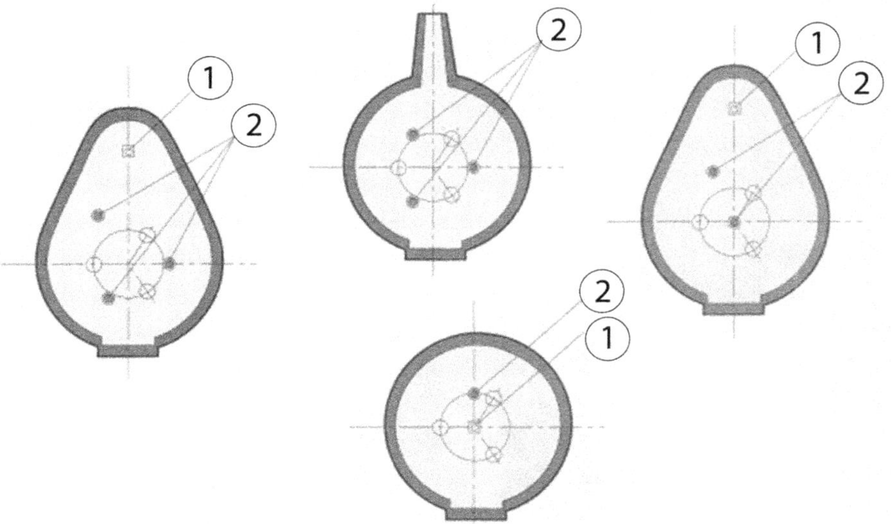

FIGURE 3.11 Diagrams of deployment of porous plugs in various designs and shapes of bottom: 1 – taphole brick, 2 – porous plugs. (Author's work.)

which is good, as it is reasonable to perform maintenance at the same time for the whole wall refractory lining. An example of wall refractory lining is shown in Figure 3.13.

The EAF wall refractory lining works in much harsher conditions than the bottom. It primarily arises from the direct, intensive radiation of electric arcs and their unequal power, erosive and corrosive impact of slag, and metal oxide vapors occurring in the furnace gaseous atmosphere.

As a result of the intensive radiation of electric arcs onto the wall refractory lining above the slag level, the so-called overheated spots occur in a significant area of the walls. In these areas on the furnace perimeter, the walls wear faster, which is caused by more difficult working conditions. The main reasons for the occurrence of overheated spots are arc plasma radiation, thrust of the arcs on the melt and a resulting blow out of slag and metal particles onto the wall, and metal bath radiation. The arc plasma thrust on the metal bath surface is so strong that the melt cannot absorb all of the thermal energy. Therefore, part of the arc plasma is reflected from the melt surface in the form of intensively shining arc beams that are directed toward the walls and at the same time "splashing" them with metal and slag. This happens, in particular, in the wall area at the shortest distance from the electrodes. The impact of the arc length and the slag layer thickness on the nature of these flames and how they impact the furnace wall is characterized by the diagram that is shown in Figure 3.14a and b. Figure 3.14a shows the case where the scrap is melting down; then, the arc flames hit the scrap and the walls are shielded from their impact. Figure 3.14b shows the case where the scrap is in the final phase of meltdown; then, the arc flames largely hit the walls that are not

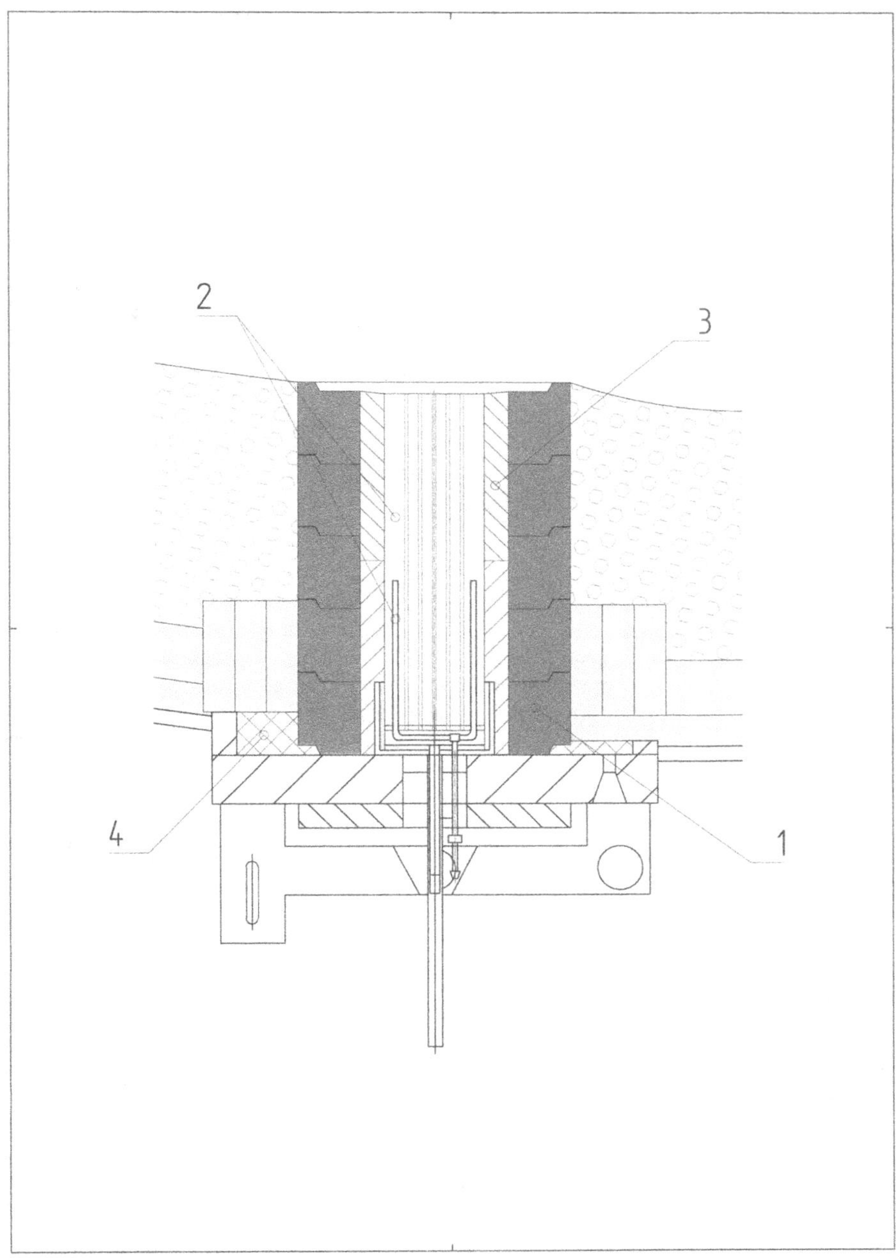

FIGURE 3.12 Structural diagram of porous plugs [6]: 1 – external ceramic bricks, 2 – internal plug with a system of thin pipes for gas flow, 3 – sealing magnesia-carbon mass, 4 – sealing of the plug in the shell. (Advertising materials (corporate brochure) of the Zaklady Magnezytowe "Ropczyce" – with permission.)

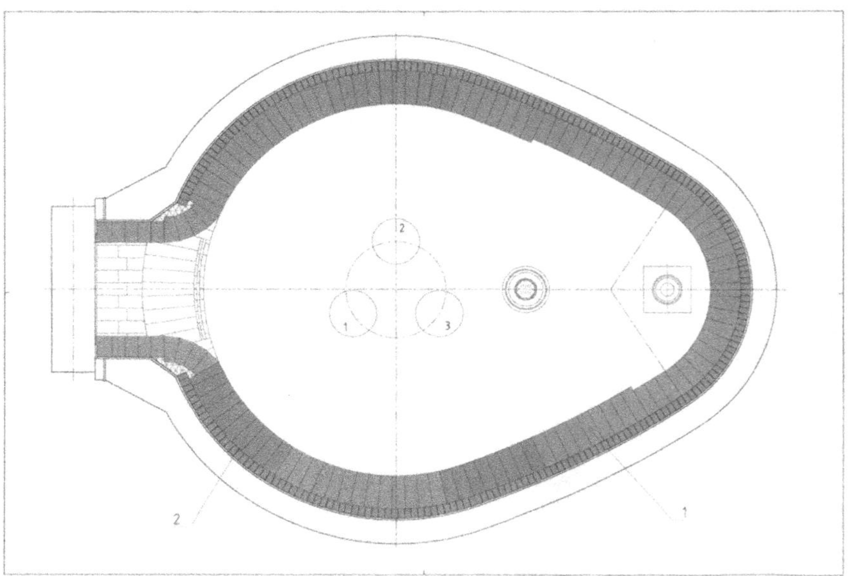

FIGURE 3.13 A horizontal cross section of an electric arc furnace at the height of the slag zone layers showing an example of the zoned installation [6]: 1 – high-quality magnesia-carbon bricks, constituting a prevailing material in the slag layer refractory lining, 2 – very high quality magnesia-carbon bricks installed in the highest wear areas, i.e. at the electrode height. (Advertising materials (corporate brochure) of the Zaklady Magnezytowe "Ropczyce" – with permission.)

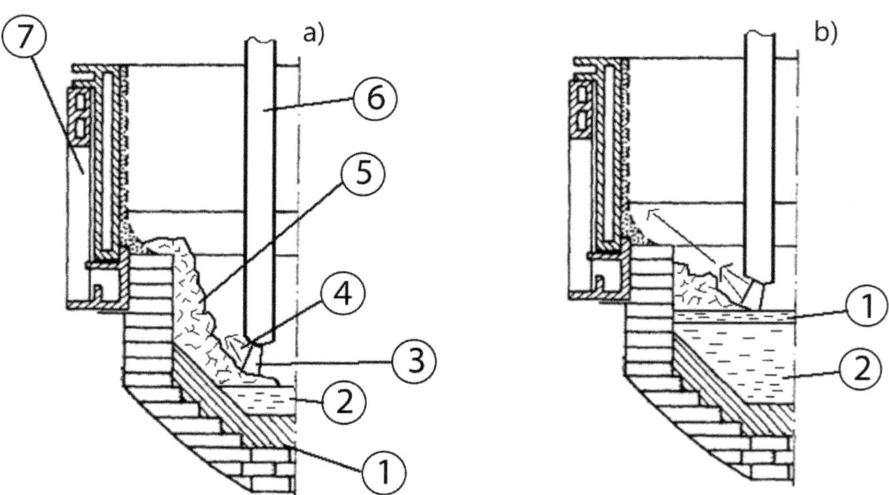

FIGURE 3.14 Diagram of arc impact on the furnace wall: (a) during the scrap meltdown, when the wall is covered, (b) after the scrap meltdown, when the wall is exposed: 1 – bottom, 2 – liquid metal, 3 – arc, 4 – arc flame, scrap, 5 – graphite electrode, 6 – walls. (Author's work.)

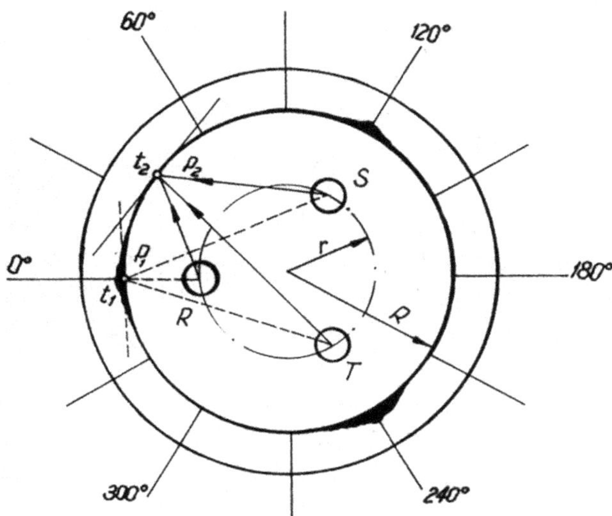

FIGURE 3.15 Geometrical conditions in the horizontal plane of electric arc radiation on the furnace wall. (Author's work.)

covered by the scrap, and so the walls are exposed to their detrimental impact. The range of this impact largely depends on the slag layer thickness. For a long arc or excessively thin slag layer, the impact of flames on the wall lining is more harmful than for a short arc and the optimal amount of slag.

In the classic arc furnace, three electrodes are deployed in the horizontal cross section of the furnace on the electrode circle at a 120° angle. As the amount of thermal energy from the arc beams reaching the wall depends on the distance, the arc impact on the wall refractory lining varies along the whole wall perimeter. The geometrical conditions of the electric arc radiation onto the furnace wall in the vertical plane are shown in Figure 3.15.

Based on experimental research results, a parameter was adopted to describe the arc energy radiation impact on the furnace walls in the form of the radiation index developed by W.E. Schwabe in the 1960s. The value of this index is proportional to the voltage and current intensity of the arc and inversely proportional to the square of the distance of the arc to the wall [7]:

$$R_w = k\frac{U^2 \cdot I}{a^2} \tag{3.4}$$

or

$$R_w = k\,\frac{(U-80)\cdot I}{a^2} \tag{3.5}$$

where:

R_w – radiation index, W/m^2,
k – unit conversion factor,
U – electric arc voltage, V,
I – electric arc current intensity, A,
a – distance from the arc to the wall, m.

Regardless of the applied formula, it appears from the index calculations that the maximum of radiation intensity occurs opposite to each electrode, whereas at half the distance between the electrodes, the radiation intensity is the lowest. The radiation intensity depends not only on the distance from the arcs to the wall but also on their power and length. On the basis of the proposed description of the arc impact on the walls (equations 3.4 or 3.5), you can quantitatively determine the linear wear of refractory materials during melting [8]:

$$R_L = k\int_0^{t_w} f(E)\frac{(U-80)\cdot I}{a^2}dt \tag{3.6}$$

where:

R_L – linear wear of the furnace walls over time, mm/min,
k – unit conversion factor taking into account the type of refractory materials (for magnesia, it is 920 mm^3/MJ; for dolomite, it is 2,600 mm^3/MJ),
$f(E)$ – function describing the dependence of the current wear on the unit energy accumulated by the charge, MJ,
t – time, min,
t_w – time of the heat, min,

other symbols as in equation (3.4).

As a result, the element of integration in equation (3.6) presents the current wear of materials during a specific heat melting. The definite integral of this element within the limits from 0 to t_w (time of melting) represents the total wear of lining during one heat due to the arc impact on the walls.

The wall refractory materials also wear because of the erosive and corrosive effect of the slag. Liquid slag in the areas of contact with refractory materials makes material constituents dissolve as a result of slag infiltration into their structure. For magnesia-carbon materials, thanks to the carbon presence, the infiltration only occurs in a few millimeters sub-surface layer. However, during operation, these products successively decarbonize; therefore, they become prone to these factors within the decarbonized area.

As demonstrated hereinabove, the working conditions of the furnace wall refractory lining are very diverse. If the lining were made in whole from a refractory material of the same quality, the material would wear in individual zones in a very uneven manner, and the unit wear of materials in the process would be high. Therefore, to improve the evenness of the wall wear, the so-called zoning is applied. This means that, in the fast wearing areas, higher quality materials are installed, with properties properly selected to match the working conditions. However, in the zones with less wear, lower quality (usually cheaper) materials are applied.

In the slag line, materials are primarily exposed to the slag corrosive and erosive impact, and to a lesser extent, rapid temperature changes. Magnesia-carbon materials show good resistance to the slag impact. They are made on the basis of high purity fused magnesia, i.e. 97%–98% MgO, with large crystals and a high total carbon content, i.e. approximately 13%–16%.

Above the slag zone, in the so-called hot spots, materials are primarily exposed to the impact of high and fast changing temperature, from the arcs and slag. The materials applied here are similar to the ones applied in the slag zone, but some steelmaking shops decide to install top chemical purity magnesia-carbon inserts in these areas.

In the upper part of the furnace walls, refractories primarily wear as a result of the impact of dust, metal oxide vapors, and temperature changes. The temperature fluctuations in this zone are lower and, therefore, the lining working conditions are much easier. Therefore, this part of the refractory lining can be made of good quality magnesia-carbon materials, albeit of lower quality than for the slag zone. As a rule, it is connected to a slightly lower chemical purity of fused magnesia used for the production of refractories, and a slightly lower content of total carbon, i.e. by approximately 1%–3%. In this area, solutions based on magnesia-chromite products are also met.

Zoned wall refractory, with correctly selected materials, good workmanship, and operation, brings about substantial financial savings for a steelmaking shop. Depending on the furnace capacity, refractory material quality, power of electric arcs, and technological and organizational conditions of furnace operation, the wall lining life is usually from 500 to 800 heats, but there are also steelmaking shops recording substantially higher results.

The optimization of the arc parameters (power, voltage and current intensity, and length) and the process conditions of the scrap meltdown (amount and type of scrap and slag, slag foaming degree) in the context of minimizing the consumption of wall refractory material are performed on the basis of the numerical modeling of the effects occurring during melting. The paper by Piero Frittell et al. [9] is one of many examples of such simulation calculations.

3.4 WATER-COOLED WALL DESIGN

As the power supply transformer capacities have increased, the operation conditions of the furnace wall and roof refractory lining have worsened. The application of water-cooling for the furnace walls and roof was one of the design solutions to improve the lining operation conditions. Thanks to this solution, the transformer capacity could be increased and, at the same time, good working conditions for the furnace walls were maintained. Furnace wall water-cooling has become a common solution in EAF design, in particular for large volume furnaces. Replacing the wall refractory lining with water-cooled elements started in the 1970s. Initially, small water-cooled panels were applied at the so-called hot spots of the wall lining in order to extend its life [10]. The method of mounting this small water-cooled panel is shown in Figure 3.16.

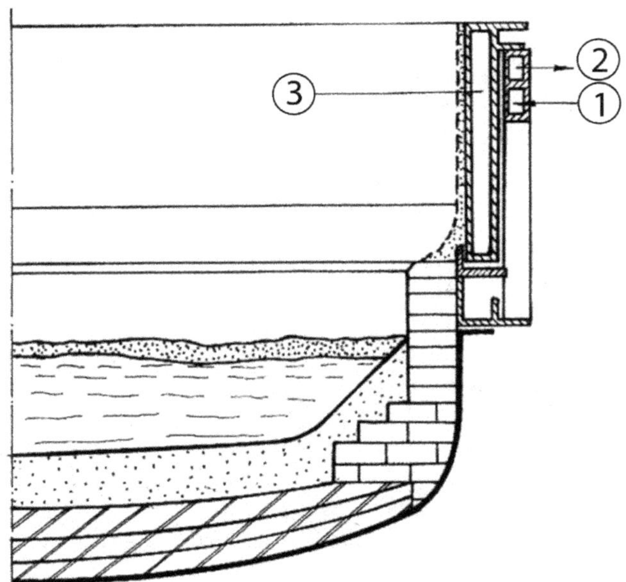

FIGURE 3.16 Example of water-cooled panel application in the hot zones of an electric arc furnace: 1 – water in, 2 – water out, 3 – water-cooled panel. (Author's work.)

In later solutions, coolers were deployed along the entire wall length. Furnace with water-cooling wall can be applied in a few design versions:

- steel plate water-cooling panels,
- layered cooling system,
- cast iron blocks with embedded water-cooling steel pipes, sometimes with embedded refractory bricks on the internal surface,
- cast steel blocks with embedded water-cooling pipes,
- water-cooling panels made of thick copper plate,
- panel cooling with steel pipes,
- system of coaxial pipes covering the furnace volume.

3.4.1 Water-Cooling Side Panel Systems

Water-cooling panels were first applied in Japanese steel plants, initially only within the hot spot zones, and in later solutions, in the form of a dozen or so segments on the entire perimeter of the furnace. This type of water-cooling may cover up to 70% of the wall side surface area. Only the wall areas immediately above the slag line, and a part of the rear wall above the taphole, are made of refractory bricks. An example of a design solution of a single water-cooling panel is shown in Figure 3.17. A water-cooling panel, made of steel plate, has the shape of a circular sector with its radius equal to the vessel radius and a height equal to the vessel height. A dozen or so water-cooling panels are installed on the vessel perimeter, with the appropriately selected

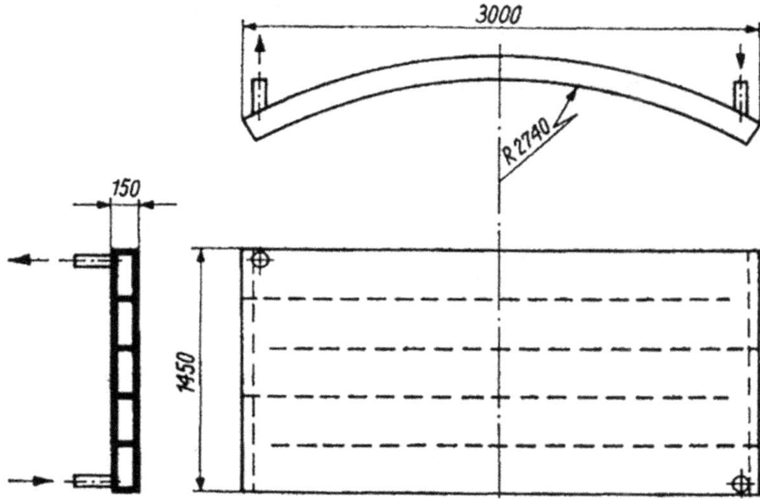

FIGURE 3.17 Example of a design solution of a single panel of water-cooled EAF walls. (Author's work.)

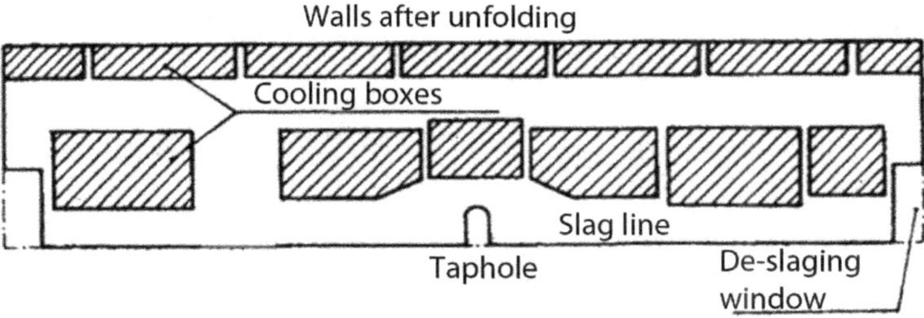

FIGURE 3.18 Layout diagram of water-cooling panels on the furnace wall perimeter. (Author's work.)

length of the circular sector. In its lower part, there is a cooling water inlet, whereas its outlet is located in the upper part.

The diagram showing the arrangement of water-cooling panels on the furnace wall perimeter is shown in Figure 3.18. Subject to the furnace capacity, 14–16 cooling panels are installed. Each of them has an individual water intake and return and a separate system for cooling the water flow rate control to ensure a uniform temperature along the furnace perimeter.

The working surface of water-cooling panels is protected with a refractory layer, usually of magnesia or magnesia-carbon materials. This lining insulates metal surfaces of panels from the impact of thermal energy from electric arcs. Figure 3.19 shows the diagram of water-cooling panel installation in the EAF wall.

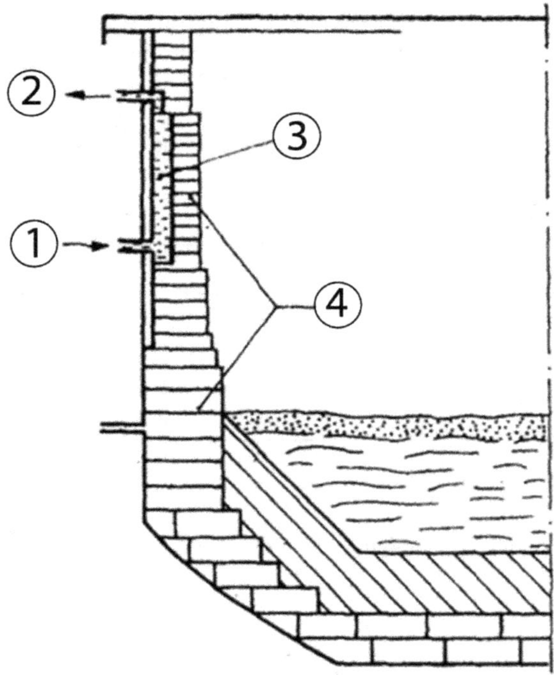

FIGURE 3.19 Diagram of water-cooling panel installation in the electric arc furnace wall with a refractory insulating layer: 1 – water in, 2 – water out, 3 – water-cooled panel, 4 – refractory bricks. (Author's work.)

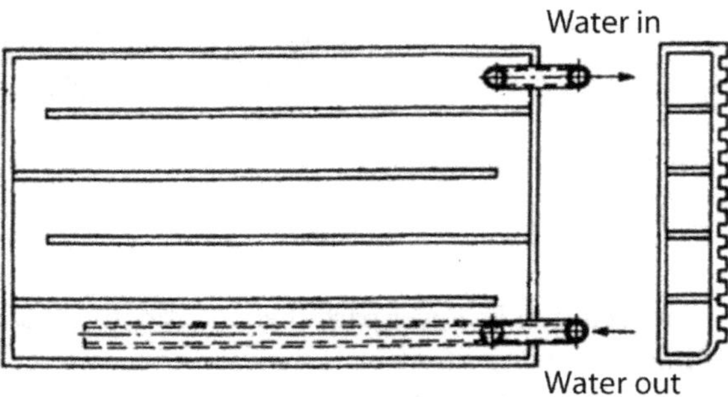

FIGURE 3.20 Example of a design solution of a water-cooling panel with a grooved internal surface. (Author's work.)

With large thermal loads, the refractory lining protecting a water-cooled panel of this design can be damaged easily. This fault has been eliminated in the design solution with a grooved internal surface of the panels. During the steel melting process, a protective slag layer builds up to insulate the steel surface of the panel from the direct impact of the electric arc radiation. An example of such solution is shown in Figure 3.20.

In large volume EAFs, the shell is split into two parts: the lower (bottom) and the upper (walls) which are cooled with water. In a solution like this, water-cooling panels can be installed in a simpler manner. Water-cooling panels are installed on the external and internal wall sides. It reduces the hazard of water leaks onto the bottom if a panel steel wall melts. The method of cooler installation with an external and internal cooling panel and solutions of the so-called safe cooling systems are shown in Figures 3.21 and 3.22.

3.4.2 Layered Cooling System

The layered cooling system of arc furnace walls is characterized by a combination of water-cooled elements and refractory materials. Welded plates box made from steel sheets are used as cooled elements. Magnesite materials, including magnesia and magnesia-carbon products, are used as refractory materials. Water-cooled panels are arranged alternately with refractories. An example of such a design solution is shown in Figure 3.23.

3.4.3 Cast Iron and Steel Water-Cooled Blocks with Cast-In Steel Pipes

Systems of cast iron and steel blocks with cast-in steel pipes are applied primarily for smaller size arc furnaces. This stems from a relatively small cooling water flow rate and a related low cooling intensity. The surface of metal cooling elements is

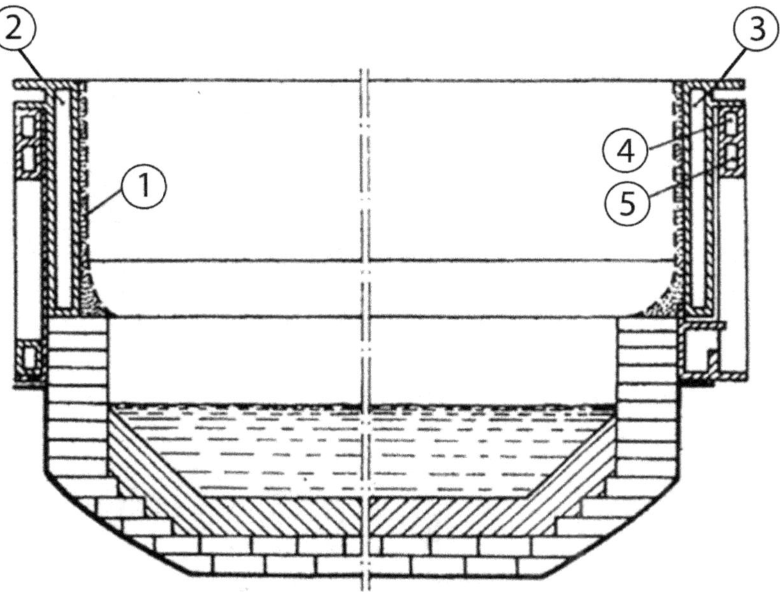

FIGURE 3.21 Design solution of a water-cooling panel installation on the outside and inside of the wall using the cage system: 1 – layer of imposed mass, 2 – box cooler inside the furnace shell, 3 – box cooler outside the furnace shell, 4 – water out, 5 – water in. (Author's work.)

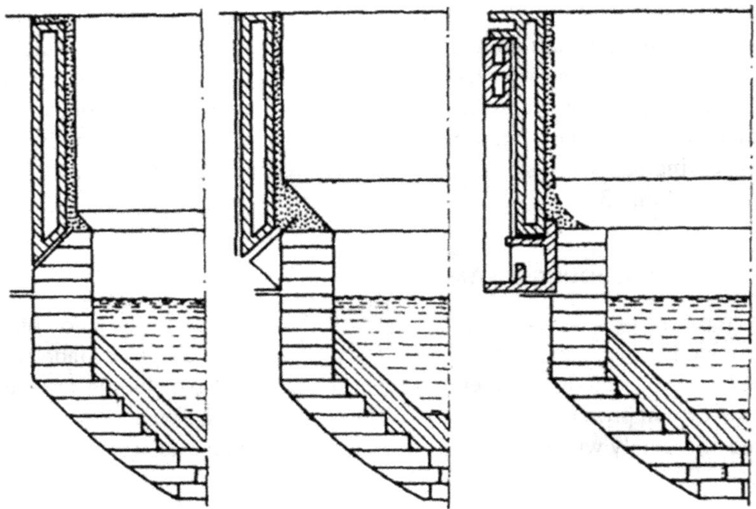

FIGURE 3.22 Examples of a design solution of the so-called safe panel cooling systems of arc furnace walls. (Author's work.)

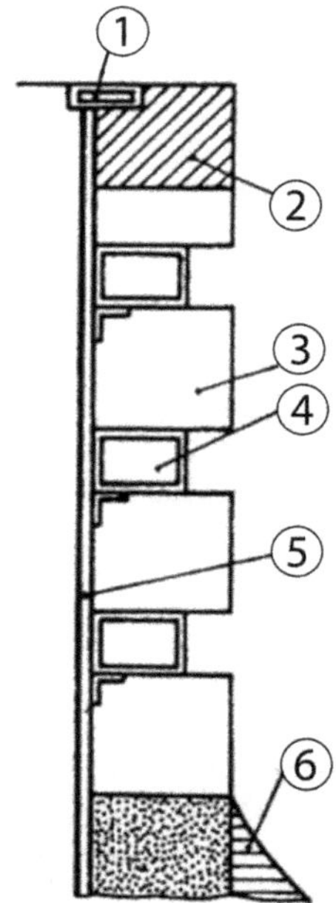

FIGURE 3.23 Design solution of a layered cooling system of EAF walls: 1 – water-cooled thrust ring, 2 – basic brick with metal cover, 3 – carbon brick, 4 – cooler, 5 – furnace shall, 6 – magnesite lining. (Author's work.)

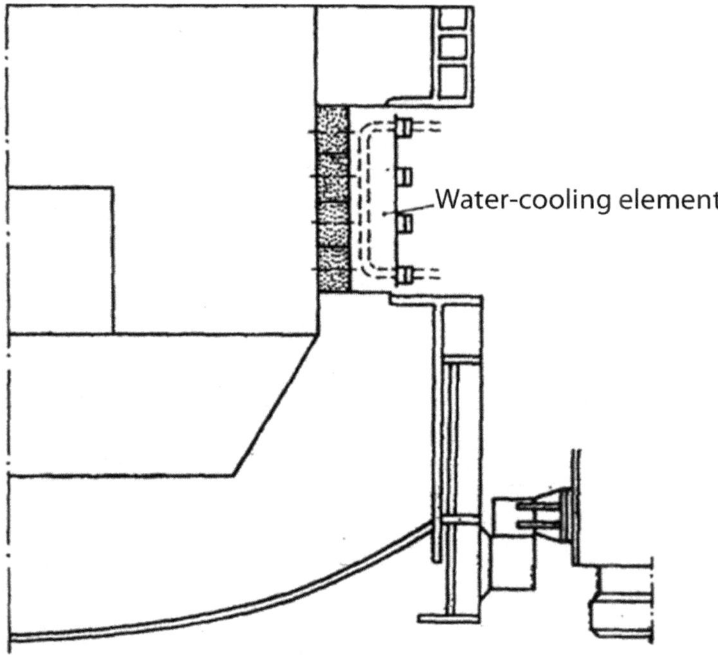

FIGURE 3.24 Design solution of cast iron or steel blocks for the water-cooling of EAF walls. (Author's work.)

usually covered with an insulating layer of magnesite materials, including magnesia, magnesia-chromite, or magnesia-carbon refractories. These systems have an advantage of no weld joints, resulting in a better resistance to the electric arc impact. An example of a design solution of such a system is shown in Figures 3.24 and 3.25.

3.4.4 Tubular Cooling System

The tubular system is the most often used design solution as regards the cooling of the EAF walls. A system like this is characterized by the most advantageous heat transfer indicator, while minimizing the so-called cold zones. Two solutions of pipe systems are applied: in a panel form and horizontal coaxial pipes encompassing the whole furnace walls perimeter. The panel system, where each panel has individually controlled cooling water intake and return, is more frequently used. Therefore, the effect of nonuniform impact of the heat flux from the electric arcs on the wall perimeter is mitigated. The cooling water flow parameters are selected in each panel so that temperatures of all panels are similar. Often metal elements are padded on the internal surface of the panels. They facilitate the formation of the slag protective layer and its build-up on the panel surface. It increases the life of cooling elements. The views of a wall water-cooling panel and a panel with padded elements for supporting the slag layer are shown in Figures 3.26 and 3.27, respectively. Individual

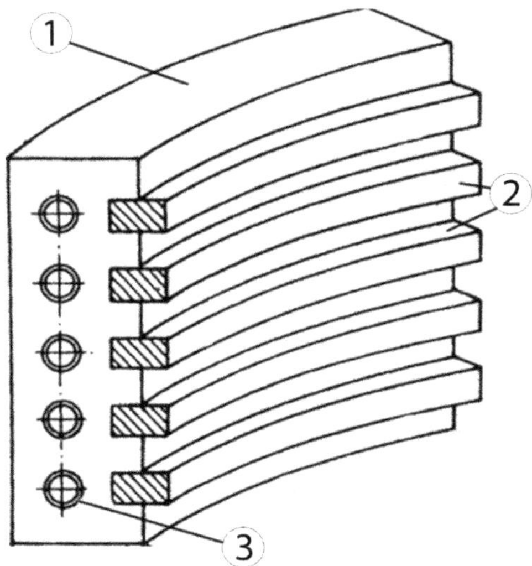

FIGURE 3.25 Diagram of a cast iron water-cooled block applied to cool the walls of an EAF: 1 – cast iron body, 2 – refractory materials, 3 – pipes for water flow. (Author's work.)

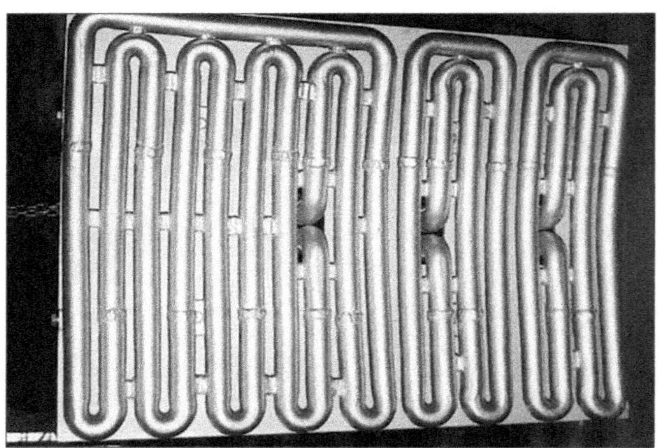

FIGURE 3.26 View of an individual panel of the electric arc furnace wall-cooling system [1]. (Advertising materials (corporate brochure) of the SMS group GmbH (formerly SMS Siemag AG) – with permission.)

panels (also called segments) are arranged along the furnace vessel perimeter, both forming the shell and replacing the wall refractory lining [10,11].

Figures 3.28 and 3.29 show views of fragments of water-cooled furnace walls where a panel with a burner and a panel above the main door can be seen on the foreground.

FIGURE 3.27 View of the EAF wall panel-cooling system with padded elements to support the slag protective layer [1]. (Advertising materials (corporate brochure) of the SMS group GmbH (formerly SMS Siemag AG) – with permission.)

FIGURE 3.28 View of water-cooled panels in the furnace along with a burner-cooling panel [1]. (Advertising materials (corporate brochure) of the SMS group GmbH (formerly SMS Siemag AG) – with permission.)

FIGURE 3.29 View of water-cooled panels in the furnace along with a panel above the main door [1]. (Advertising materials (corporate brochure) of the SMS group GmbH (formerly SMS Siemag AG) – with permission.)

3.4.5 Principles of Cooling System Engineering

Designing EAF water-cooling systems is based on the density of heat flux reaching the wall surface. According to Kepler's law, the heat flux coming from a source with power P to any point has a density of [2]:

$$q = \frac{P \cdot \cos\alpha}{4\pi r^2} \tag{3.7}$$

where:

q – heat flux density at a certain point in the space, W/m^2,
P – power of a point heat source, W,
α – angle between a straight line going through the heat source and the point at which the heat flux is determined, and the normal plane at this point (the tangent plane to the surface in which this point is located),
r – distance between the heat source and the point at which the heat flux is determined, m.

For an EAF there are three heat sources (three arcings), from which the thermal energy reaches any point at the furnace wall surface. The total heat flux, at each point of the wall surface, is the sum of three fluxes coming from the three arcs. The diagram of the geometrical distribution of heat fluxes is shown in Figure 3.30.

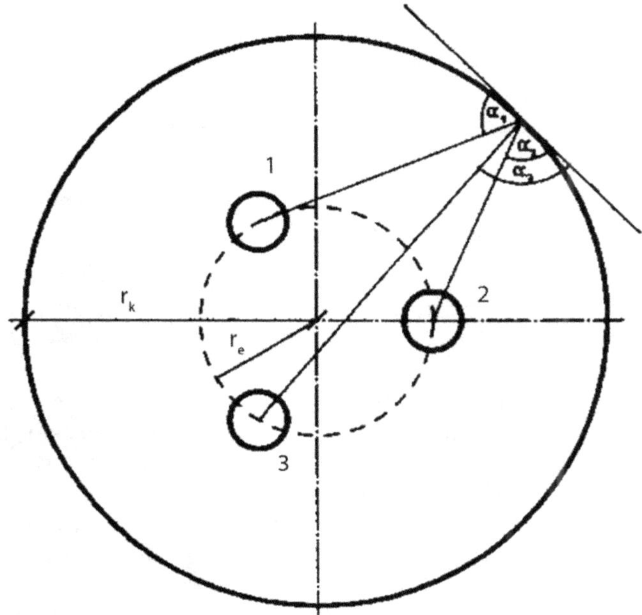

FIGURE 3.30 Diagram of an electric arc furnace section with plotted angles for computing the heat flux at any point of the walls: 1, 2, 3 – subsequent electrodes. (Author's work.)

The Kepler law for a case like this can be written as follows: [12–14]:

$$q = \frac{1}{4\pi r_e^3} \sum_{i=1}^{3} \frac{P_i \left(1 - \frac{r_e}{r_k} \cos\alpha_i \right)}{\left[1 + \left(\frac{r_e}{r_k} \right)^2 - 2\frac{r_e}{r_k}\cos\alpha_i + \frac{(H-h)^2}{r_k^2} \right]^{\frac{3}{2}}} \tag{3.8}$$

where:

- q – heat flux coming from the three arcs at any point of the furnace wall, kW/m^2,
- P_i – active power of the arcs, kW,
- r_e – electrode pitch circle radius, m,
- r_k – metal bath surface radius, m,
- α_i – angle between the straight line going through the axis of a subsequent electrode "i" and the point at the wall furnace for which the heat flux is calculated and the tangent plane to the furnace wall at this point (Figure 3.31),
- H – distance from the point at which the heat flux is calculated to the metal bath surface, m,
- h – length of the arc not covered by the slag, m.

The analysis of equation (3.8) indicates that the heat flux distribution around the walls is nonuniform. It achieves its maximum value at the points closest to the subsequent three electrodes. However, its lowest value is achieved at the points with the same distance to the two neighboring electrodes. At the same time, the values of heat flux depend on the ratio of the electrode pitch circle to the radius of metal bath surface. The value of this ratio should be selected to minimize the nonuniformity of the heat flux distribution. To this end, the minimum of the function being the quotient of the maximum and minimum flux relative to $\frac{r_e}{r_k}$ is determined. This condition is used to establish the value of electrode pitch circle diameter for which the nonuniformity of the distribution of heat flux from the electric arcs on the furnace walls is the lowest, knowing the furnace diameter at the melt level.

The design parameters of the cooling system are adjusted to the highest value of the computed heat flux. When designing this system, almost all its geometrical and material-related parameters are assumed, and the next one checks if the boundary conditions are met such as the outlet water temperature, amount of steam generated, and temperature of cooler elements. The pipe diameter and the type of pipe material, flow velocity, inlet water pressure and temperature, and the heat flux are assumed. Based on these assumptions, the pipe length in one segment of the cooling system is determined, for the appropriate outlet water parameters.

In the general perspective, there is a relationship between the heat flux and the temperature gradient, described with the Fourier's heat conduction law [15]:

$$q = -\lambda \frac{dT}{dx} \tag{3.9}$$

where:

q – heat flux density, W/m^2,

λ – specific heat conductivity, W/m·K,

$\frac{dT}{dx}$ – temperature increment over distance x, K/m.

The minus sign in equation (3.9) means that heat flows in the direction opposite to the existing temperature gradient. The Fourier law concerns heat transfer by conduction. It means that the heat flux flowing through the pipe wall with a thickness $g = r_2 - r_1$ will have the density of:

$$q_r = \frac{\lambda}{r \ln \frac{r_2}{r_1}} (T_2 - T_1) \tag{3.10}$$

where:

q_r – density of heat flux flowing through the pipe, W/m^2,

λ – specific heat conductivity of the material of which the plate or pipe is made, W/m·K,

r_1, r_2 – pipe radius, internal and external, respectively, m,

r – current pipe radius, m,

T_1, T_2 – temperature of the pipe external or internal surface, *K*; index 1 refers to the lower temperature, and index 2 refers to the higher one.

Heat conductivity coefficient λ, for instance, for steel is 60 W/m·K and for copper is 380 W/m·K. If heat flows between a solid body and the adjoining liquid, it flows by convection. In this case, the effect of convective flow is described mathematically by the Newton equation [15]:

$$q = \alpha(T_1 - T_2) \tag{3.11}$$

where:

q – heat flux density, W/m^2,
α – heat transfer coefficient, W/m^2·K,
T_1, T_2 – temperature of the metal surface from or to which the heat flows and the temperature of the flowing liquid, respectively, K.

The heat transfer coefficient α can be determined for various flow states and for various geometries of the system on the basis of nondimensional criteria numbers [15,16].

When designing an EAF water-cooling system, two fundamental principles should be considered:

- strive to achieve heat transfer that is as good as possible, with internal flow in tubular elements,
- the flow parameters should be selected so as to avoid exceeding the critical load of the cooler surface.

In practice, the following types of EAF cooling systems are applied, and they only differ with the temperature of the flowing water:

- Cold cooling: the water flowing into the cooling elements has the ambient temperature, and the outflowing water temperature does not exceed 313 K (40°C),
- Hot cooling: the water flowing into the cooling elements has a temperature of approximately 50°C below the boiling point, and the outflowing water temperature also does not exceed the boiling point,
- Evaporative cooling: the water flowing into the cooling elements has a temperature slightly below the boiling point, and a mixture of boiling water and steam flows out.

In the case of cold cooling, the cooling water is not used for other purposes. For hot cooling, it is used as hot tap water, and for evaporative cooling, it can be utilized to generate steam.

Computing methods for water-cooling systems are based upon the energy balance. For a system with tubular elements containing flowing cooling water, with a constant unit heat flux supplied, and without water evaporation, the energy balance can be written as follows:

$$q\pi Dl = mc_w \left(T_2 - T_1\right) \tag{3.12}$$

where:

q – density of the supplied heat flux, W/m^2,
D – inner pipe diameter (water-cooling element), m,
l – pipe length (water-cooling element), m,
m – overall mass flow of the cooling water, kg/s,
c_w – specific heat of the cooling water, J/kg·K,
T_2, T_1 – cooling water temperature at the pipe outlet and inlet, respectively, K.

After the appropriate transformations of equation (3.12), and taking into account pipe geometrical relationships and the flow rate of the cooling water, an equation describing the temperature of the water flowing in the system as a function of length is as follows:

$$T_2(l) = T_1 + \frac{4qd_z}{\pi\rho_w c_w w d_w^2} l \tag{3.13}$$

where:

q – density of the supplied heat flux, W/m^2,
d_z, d_w – external and internal pipe diameter, respectively, m,
ρ_w – specific mass of the cooling water, kg/m^3,
w – flow velocity of the cooling water, m/s,
c_w – specific heat of the cooling water, J/kg·K,
l – distance for which the T_2 is calculated, m,
T_1, T_2 – cooling water temperature at the pipe inlet and at the "l" length of the pipe, respectively, K.

Such a linear growth in the cooling water temperature as a function of distance is observed until the temperature T_2 reaches the value of water boiling point. The water boiling temperature depends on the pressure in the system. The distance from the beginning of the cooling element, at which boiling occurs, can be calculated from the transformed formula (3.10), for $T_2\ (l) = T_w$. Changes in the temperature of the cooling element external surface can be determined from the transformed equation (3.11).

The relationships derived hereinabove are correct until boiling starts in the system. Then, in a certain section of the cooling element, the cooling water temperature is constant and equal to the boiling temperature. The length of the cooling element at which boiling occurs can also be determined.

The above guidelines and assumptions are presented, enabling a water-cooling system of the EAF walls to be designed, including the correct selection of geometrical and material-related parameters of coolers, and parameters of the flowing water for the assumed thermal conditions inside the furnace. More details about designing the EAF wall water-cooling system can be found in the references [2].

3.5 DESIGN AND STRUCTURE OF THE ROOF

The roof of an EAF is its independent structural component, not permanently connected to the furnace. Thanks to this solution, the roof can be quickly lifted up in order to load the charge and to be easily replaced during maintenance. Typically, for smaller size furnaces, the roof is made of refractory materials. Modern furnaces have water-cooled roofs (similar to the furnace walls). The roof is dome shaped, and it is supported by a steel roof ring. The roof ring is made of welded steel plates, and usually it is water cooled. Various design solutions of roof rings are shown in Figure 3.31. In order to increase the contact area of water and the steel plate surface, special internal partitions are applied. Sometimes the ring is made of steel tubes. This ring design enables the cooling water to be supplied independently.

The furnace roof is suspended from a semi-gantry structural frame at four points, with the use of welded hooks. The hooks are used both for the suspension during operation and for transport. The structural frame consists of two cantilever arms, of the jib type, topped with a crosswise beam. The roof together with the load-carrying gantry is secured to a movable, vertical structural column. A service platform for handling electrodes can be installed on the cantilever jibs. During furnace charging, the roof must have the capability of being lifted and swung outside the furnace contour. Two hydraulic cylinders are used to move the roof: one of them is for lifting, and the other one for driving the roof swinging mechanism. The lifting height is 200–400 mm, and the swinging angle, most often toward the tapping side, is 80°–100°. In the bottom part of the roof ring, there is a flanged seal of the top part of the furnace walls.

In the central part of the roof, there are three holes in the refractory lining for inserting graphite electrodes. To prevent sticking electrodes in the roof holes during operation, the hole diameters are 20–40 mm larger than the electrode diameters. A stream of hot, dusty gases escapes through a gap formed this way, which considerably worsens the operation conditions of electrode holders.

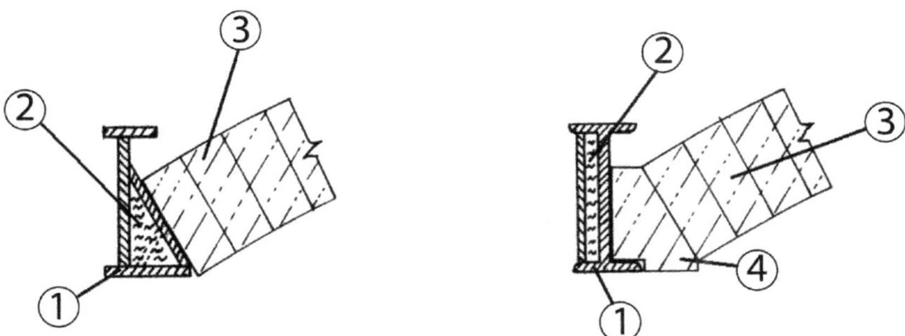

FIGURE 3.31 Potential design solutions of the roof ring [3]: 1 – vault ring, 2 – cooling water, 3 – refractory lining, 4 – retaining fitting. (Source: Charles R. Taylor 1985 – with permission.)

There is also a fourth hole in the roof, and it is used to evacuate gases formed during the steel melting process, the so-called off-gases. Gases are evacuated through a special furnace elbow structure, secured to the roof "at the fourth hole." The elbow, being an integral part of the roof, is made of steel, water-cooled components, and it is the initial element of the gas exhaust system. In the normal position of the furnace vessel and the roof, the elbow outlet goes directly to the further part of the off-gas exhaust and cleaning system. Sometimes additional holes are applied in the roof for putting charge materials to the furnace, such as lime or direct reduction materials.

The roof lining is nowadays primarily made of high-alumina bricks, which are classified in the group of aluminosilicate materials [4]. The materials are made from fireclay, kaolin, natural anhydrous aluminum silicate, technically pure aluminum oxide, sintered corundum, molten alundum, and mullite. Mineral kaolinite ($Al_2O_3{\cdot}2SiO_2{\cdot}H_2O$) is the main component of fireclays. After calcination, it contains 46% Al_2O_3. By adding calcinated raw materials with a high Al_2O_3 content to fireclays, products with an increased alumina content, which are also called high alumina materials, can be obtained. These materials most often contain 60%–80% Al_2O_3. Materials containing about 72% Al_2O_3, which corresponds approximately to the theoretical share of aluminum oxide in mullite ($3Al_2O_3{\cdot}2SiO_2$), are called mullite materials.

High alumina roof materials primarily wear out as a result of the reactions of ferrous oxides and calcium oxide with aluminum oxide and mullite that are contained in the material, thereby causing structural changes. The type and amount of the forming products of reactions depend on the concentration of ferrous oxides and calcium oxides in the furnace gaseous atmosphere and the properties of high alumina materials. The following are the basic products of the reaction with CaO: anorthite ($CaO{\cdot}Al_2O_3{\cdot}2SiO_2$) and calcium hexaaluminate ($CaO{\cdot}6Al_2O_3$). Anorthite or hexaaluminate is the prevailing phase, depending on the initial SiO_2 content in the material and the amount of absorbed CaO. When the amount of ferrous oxides is adequate, the phase ($CaO{\cdot}Al_2O_3{\cdot}2Fe_2O_3$) forms, and when alkalis are present, kaliophilite and nepheline form.

Spinels are the basic products of reactions with ferrous oxides, and their type depends on the chemical composition of dust contained in the gaseous atmosphere of the furnace. If FeO prevails in the dust, hercynite ($FeO{\cdot}Al_2O_3$) forms. For a high content of FeO in the dust, fayalite ($2FeO{\cdot}SiO_2$) and ferrous cordierite ($2FeO{\cdot}Al_2O_3{\cdot}5SiO_2$) also form. When conditions are favorable for the formation of Fe_2O_3, the phase ($CaO{\cdot}Al_2O_3{\cdot}2Fe_2O_3$) appears in the working layer of the roof.

Ferrous oxides penetrate the material deeper than CaO, and they primarily determine the chemical composition of products of reactions of the refractory material with dusts. It is only when the ferrous oxide content in the furnace atmosphere particulates is insufficient that the chemical composition of the forming reaction products in the roof working layer will be determined by CaO. To obtain a satisfactory life of high alumina roofs, it is desirable to have the optimal ratio Al_2O_3/SiO_2 in the material, as at this ratio minimal amounts of anorthite form as well as excessive CaO cannot react with Al_2O_3 in the working layer. Increasing the Al_2O_3 content in the material is one of methods for restricting the amount of the forming anorthite. The above-discussed chemical reactions result in the spalling of the working surface of the roof materials and a systematic decrease in their thickness. The life of high alumina roofs is 50–140 heats, and it depends on the furnace capacity, the range of steels manufactured, and the quality of used materials.

The classic method of brick arrangement applied during a high alumina roof laying is shown in Figure 3.32. Now, monolithic roofs made of refractory concrete are often employed. Then, the roof is a single element cast from high alumina masses, with formed holes for electrodes and gas off-gas evacuation. This method of roof making has considerably improved its useful life and reduced its preparation time [11].

Similar to the walls, a modern roof is made from water-cooled elements in either a tubular or panel system, with an individual water supply. The introduction of this method stemmed from positive former results of water-cooling for walls. The first design solutions included water-cooling only for a small area of a part of the roof near the roof ring. Now, thanks to the design development about 85% of the roof area can be cooled. Only the other 15% of the area, in the direct vicinity of electrodes, is still made of refractories. High alumina, so-called refractory concretes, consisting

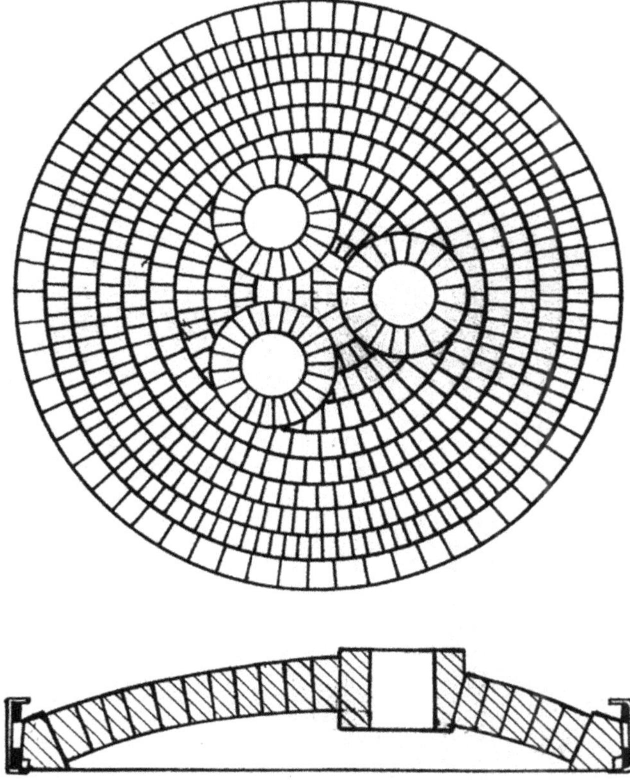

FIGURE 3.32 The classic method of brick arrangement employed during high alumina roof laying. (Author's work.)

mainly of Al_2O_3, are used for this purpose. These materials protect against a potential short circuit in the case when a graphite electrode (live) touches the metal roof. The water-cooled roof, similar to the traditional one, is dome shaped.

The roof, like the side walls of an EAF, is exposed to the impact of arc thermal energy. In addition, metal water coolers of the roof have a problem related with a hazard of inducing eddy currents and potential short circuit between an electrode and a metal element. The initial scrap melting period is particularly dangerous when the arcs glow close to the roof. The cooling water flow rate applied in industrial practice is approximately 70 L/min m^2, and the temperature difference between the outlet and the inlet is on average approximately 12°C. Compared to the traditional roof, no additional electricity consumption has been observed. The water-cooled roof service life is ten times higher than that of the traditional furnace roof made entirely from refractory materials.

In practice, two designs of roof water-cooling systems are the most common:

- systems with a panel design of cooling elements,
- systems with tubular cooling elements.

A schematic layout of individual components of the EAF roof is shown in Figure 3.33. The layout of electrode holes corresponds to their arrangement in the so-called electrode pitch circle (a hypothetical circle on which the electrodes are placed). It is a furnace design parameter expressing the diameter of a circle, which depends on the furnace size, the installed electric power of the furnace transformer, and the diameter of graphite

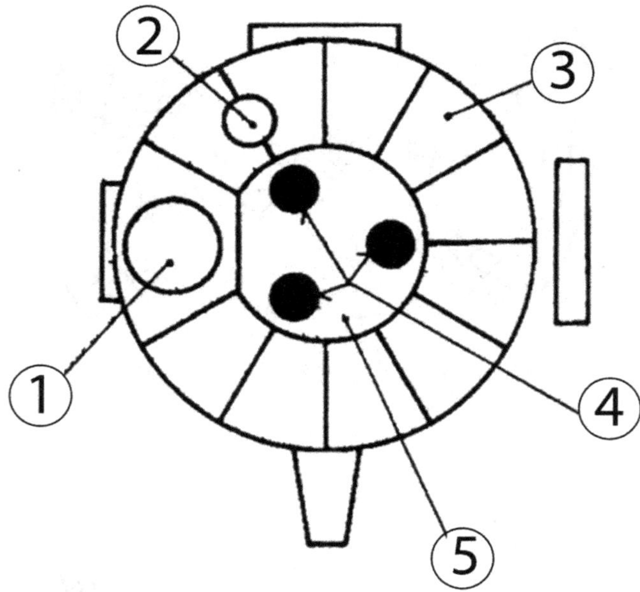

FIGURE 3.33 Layout diagram of individual components of the EAF roof: 1 – hole for off-gas capture, 2 – hole for feeding additives to the furnace, 3 – water-cooled segments, 4 – electrodes, 5 – middle part made of refractory materials. (Author's work.)

electrodes. The distance angle on the electrode circle between all holes is the same and it is 120°. Sometimes it is said that the electrode holes are placed at the vertexes of a hypothetical equilateral triangle, on which the electrode pitch circle is circumscribed. In addition, a hole for exhausting off-gases and dust is present in the structure of each roof [11].

Many design solutions of roofs with panel-type water-cooled elements are applied. The layout shown in Figure 3.34 is used more often. In this solution, the peripheral part of the roof surface is constructed as panel-type water-cooled elements. The whole structure is supported by an external ring, welded from steel plates, as in the case of classic refractory roofs (Figure 3.31). This ring usually has a hollow structure with internal partitions to increase the contact with cooling water. Hollow panels, also welded from steel plates, are secured to the internal ring wall on the roof perimeter. The shape of individual panels is adjusted to the "dome" shape of the roof, with a longer external side and a shorter internal side. They also have special steel partitions to increase contact with water. A dozen or so panels with individual cooling water supply and discharge are placed on the circumference. The panel in which the furnace elbow for off-gas discharge is secured has a slightly different design. The central part of the roof, in the part with electrode holes, is made of refractory ceramic bricks or in the form of the so-called small roof of refractory concrete. The method of manufacturing and the employed materials are analogous as in the case of a classic "full" roof.

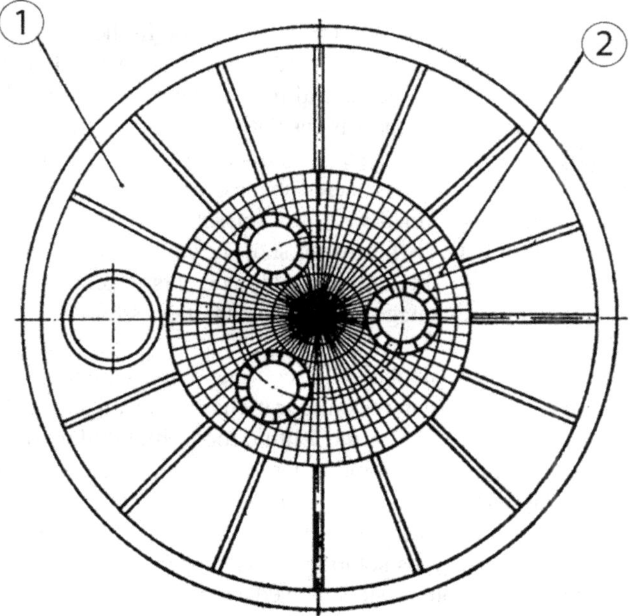

FIGURE 3.34 Design solution of the roof with the peripheral panel-type part cooled with water and the internal part around the electrodes made of refractory materials: 1 – middle part made of refractory materials, 2 – water-cooled segments. (Author's work.)

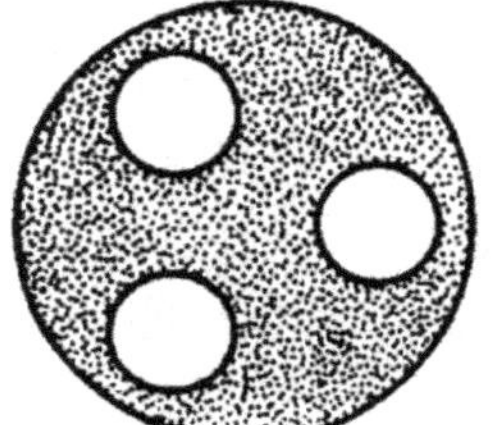 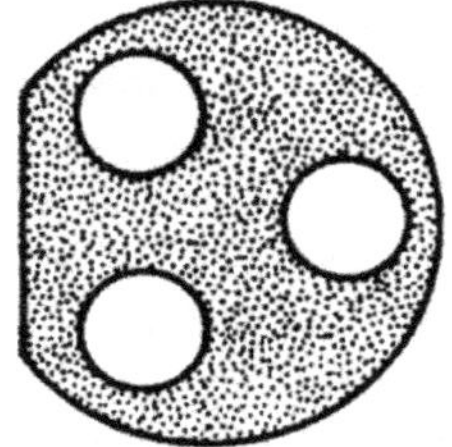 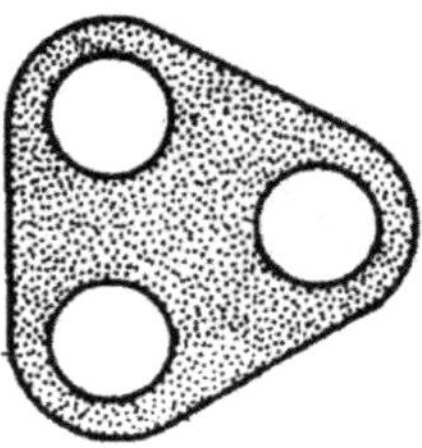

FIGURE 3.35 Examples of the shape of central part of the roof, made of refractory materials. (Author's work.)

Various shapes of the central roof part around electrodes are applied. Examples are shown in Figure 3.35. The round shape, presented in the left-hand part of the drawing, is classic. The central example illustrates an "incision" near the hole for off-gas and dust evacuation. The shape in the right-hand part of the drawing is used to minimize the amount of refractories applied. In this case, the maximum part of the roof surface is cooled with water.

Another example of a solution is the roof design with almost the entire surface cooled with water, as shown in Figure 3.36. The structure and the panel arrangement in the peripheral part of the roof are constructed in the same way as in the previous case. However, the central part of the roof is also made in the form of water-cooled hollow panels from steel plates. Three panels are made, around each of the electrodes, with a shape adjusted to the central part of the roof and with holes for electrodes. Each panel of the central part is insulated from the others, and from panels of the peripheral part, with a thin layer of refractories. Ceramic bricks of high alumina materials, with appropriately selected shapes, are most often used as insulation. Similarly, electrode holes are lined with thin refractory layers. All these insulating layers are applied to mitigate the hazard of short circuits between the electrode and panel as well as to eliminate the effect of eddy currents generated in metal panels. Such a structure has a disadvantage of a low mechanical strength and no resistance to mechanical and thermal shocks.

In recent years, roofs with tubular water-cooling systems are more and more often employed. Steel pipes with various configurations and arrangements are the main component of a structure like this. A monolithic roof with tubular cooling, as shown in Figure 3.37, is an example of an older design. The peripheral part of the roof is supported on the roof ring, and it is made of steel pipes placed radially in the form of a single system covering the roof surface. The central part in this solution is made of refractories, as in the previous solutions. It is supported on an additional water-cooled back-up ring secured on the steel pipe edges. In this solution, one water supply and one water discharge are made for the whole system. The cooling pipe diameter varies along the roof radius. In this solution, it is difficult to maintain a fairly equal temperature of the whole roof surface. It has an advantage of a simple design of cooling water supply and discharge to the whole system.

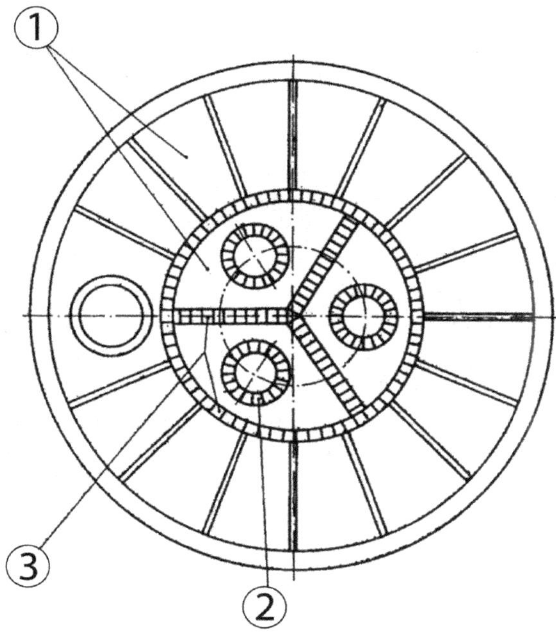

FIGURE 3.36 The design solution of virtually the whole roof in the form of water-cooled panels with insulating refractory elements: 1 – water-cooled segments, 2 – water-cooled segments. (Author's work.)

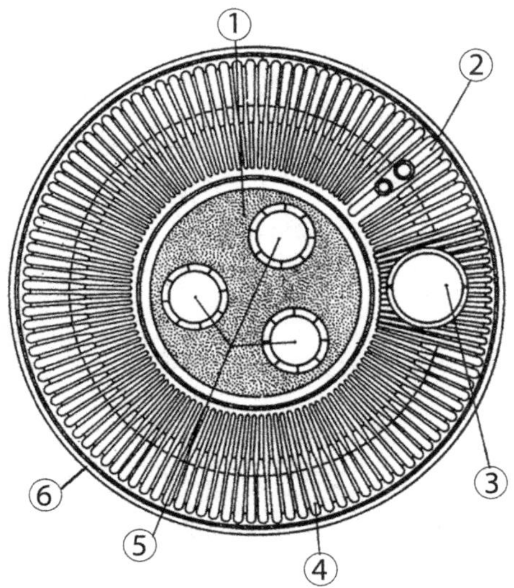

FIGURE 3.37 Diagram of a water-cooled tubular roof with a monolithic design: 1 – refractory materials around the electrode holes, 2 – water in and water out, 3 – off-gas capture hole, 4 – water-cooled segments, 5 – electrode holes, 6 – support ring. (Author's work.)

A water-cooled roof in a system consisting of a few or more panels arranged on the roof surface is another solution. Individual panels are made of adjacent steel pipes, through which the cooling water flows. An example of a solution like this is shown in Figure 3.38. The structural array of pipes in a single panel is shown at the bottom of the picture. The central part in this solution is made analogous as in the previous solution. In this system, cooling water is supplied and discharged for each panel individually. The cooling pipe diameter is the same along the whole length in the panel. A solution like this ensures that it is easy to maintain a uniform temperature of the whole roof surface. However, the cooling water supply and discharge system is more complicated.

The bottom view of a water-cooled roof after swinging from the operating EAF is shown in Figure 3.39. The picture shows the heated up tips of graphite electrodes protruding from the roof holes. In the right-hand side bottom view, you can see an off-gas exhaust hole and the furnace elbow secured to the top part of the roof above the hole. The whole internal roof surface is dark, which means that its temperature is fairly uniform and relatively low. This surface is covered with a thin layer of slag, formed from dust build-up during melting.

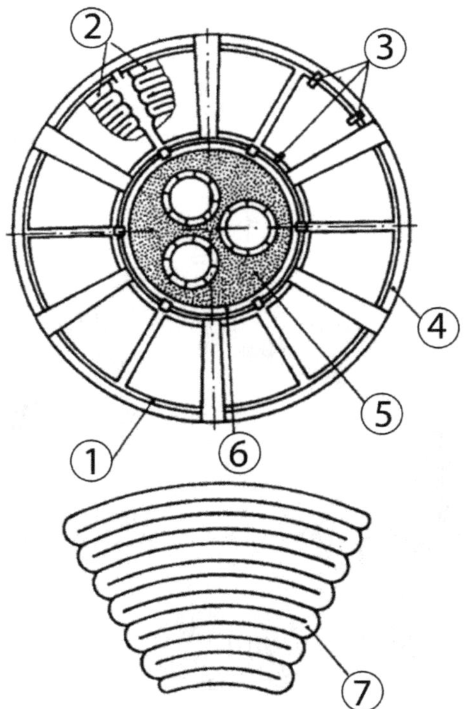

FIGURE 3.38 Diagram of a water-cooled tubular roof with a panel design: 1 – support ring, 2 – water-cooled segments, 3 – fastening of a water-cooled element, 4 – outer thrust ring, 5 – middle part made of refractory materials, 6 – inner thrust ring, 7 – tubular water-cooled components. (Author's work.)

FIGURE 3.39 The bottom view of a water-cooled roof after swinging from the operating electric arc furnace [17]. (Advertising materials (corporate brochure) of the SMS group GmbH (formerly Mannesmann Demag Huttentechnik SMS Siemag AG) – with permission.)

Figures 3.40 and 3.41 show accordingly bottom and top views of the water-cooled roofs with the tubular panel design, for smaller and larger size furnaces, with either a round or triangular central part. You can see a tubular roof ring on the outside, and the panels, also tubular, with the cooling water supply and discharge system. The central part contains a ring, also tubular, water cooled, where a separately made refractory roof component is placed.

Figure 3.42 shows a side view of the furnace elbow, placed on the roof hole. As you can see, the elbow is made of steel pipes, also water cooled.

3.6 MECHANICAL EQUIPMENT

EAFs are usually built to be tiltable. A furnace is mounted on a foundation, whose design is adjusted to the loads transferred onto the base. The foundation is made of reinforced concrete, and its shape is adjusted to the method of furnace loading and operation. Steel rails are fixed to the so-called plinth, which is the top surface of the foundation. Depending on the furnace tilting manner, various designs of the rail surface are applied – flat or rack type. The furnace rocker is supported on the rail surface, and it transfers loads from the furnace weight, including the roof, electrodes, and all of the drive gears. The foundation comprising two bases with a passage between them, in which a track for the slag car is laid, is the most often used solution.

FIGURE 3.40 The bottom view of the water-cooled roof for a smaller size electric arc furnace, tubular panel type, with a round central part [17]. (Advertising materials (corporate brochure) of the SMS group GmbH (formerly Mannesmann Demag Huttentechnik SMS Siemag AG) – with permission.)

FIGURE 3.41 The top view of the water-cooled roof for a larger size electric arc furnace, tubular panel type, with a triangular central part [17]. (Advertising materials (corporate brochure) of the SMS group GmbH (formerly Mannesmann Demag Huttentechnik SMS Siemag AG) – with permission.)

In addition, the foundation has chambers, where the furnace tilting mechanism drive and brackets for the roof swinging mechanism drive are situated.

The furnace tilting mechanism is its important structural component. It enables the furnace to be tilted to the tapping side at an angle depending on the tapping method: 40°–42° for a pouring spout and 10°–15° for a taphole in the bottom. The tilting angle toward the main door is 10°–12°; this angle should ensure the possibility of slag skimming and convenient temperature measurements, and other process operations for the operators. The tilting rate for smaller size furnaces is approximately 1.5°/s, and for medium and larger size furnaces, it is between 0.4° and 0.8°/s. During

FIGURE 3.42 The view of a furnace elbow for exhausting off-gases, fixed on the roof hole [17]. (Advertising materials (corporate brochure) of the SMS group GmbH (formerly Mannesmann Demag Huttentechnik SMS Siemag AG) – with permission.)

tilting, the furnace must retain its stability and, therefore, this operation should be performed smoothly, at a specific tilting rate. Tiling mechanisms should feature:

- simple design, easy operation, and long life,
- easy and smooth control of the furnace tilting rate,
- an inclination of the tapping spout end in the vertical plane that is as small as possible for the maximum furnace tilt,
- location preventing liquid metal or slag being poured during their potential "escape" from the furnace.

The furnace tilting mechanism includes the rocker rails resting on the foundation, the two rockers connected together with a steel frame, and the drive. The flat rocker rails resting on the foundation are cast from iron. This solution is applied in particular for medium- and larger size furnaces. The rockers are made of a steel structure, in the "cradle" shape, and they rest directly on the rails. To prevent the rocker from sliding against the rails, both elements should be dovetailed.

Either hydraulic or, less frequently, electric drive can be used to tilt the furnace. Regardless of the drive type, it is made symmetrically, i.e. each rocker has its own drive. If one rocker drive fails, then the furnace can be tilted with the other rocker drive only. The hydraulic cylinder axis inclination angle in relation to the furnace vertical axis is approximately 25°. When tilting, the furnace toward the tapping spout or toward the main door, the point of application of the force to the rocker moves on a cycloid, which guarantees its smooth movement. In such drives, it is easy to ensure smooth movement velocity control, which enables the smooth tilting of the furnace. The view of a furnace tilting mechanism rocker with a hydraulic drive from the side and from the top is shown in Figure 3.43.

For smaller size furnaces, guides in the form of rollers located on the foundation on which the furnace is mounted are applied (Figure 3.44). A solution like this enables the more stable movement of the vessel against the foundation.

Depending on the furnace type, in the rocker frame, apart from the furnace vessel, there is a mast for roof suspension, electrode holding system stands, roof and

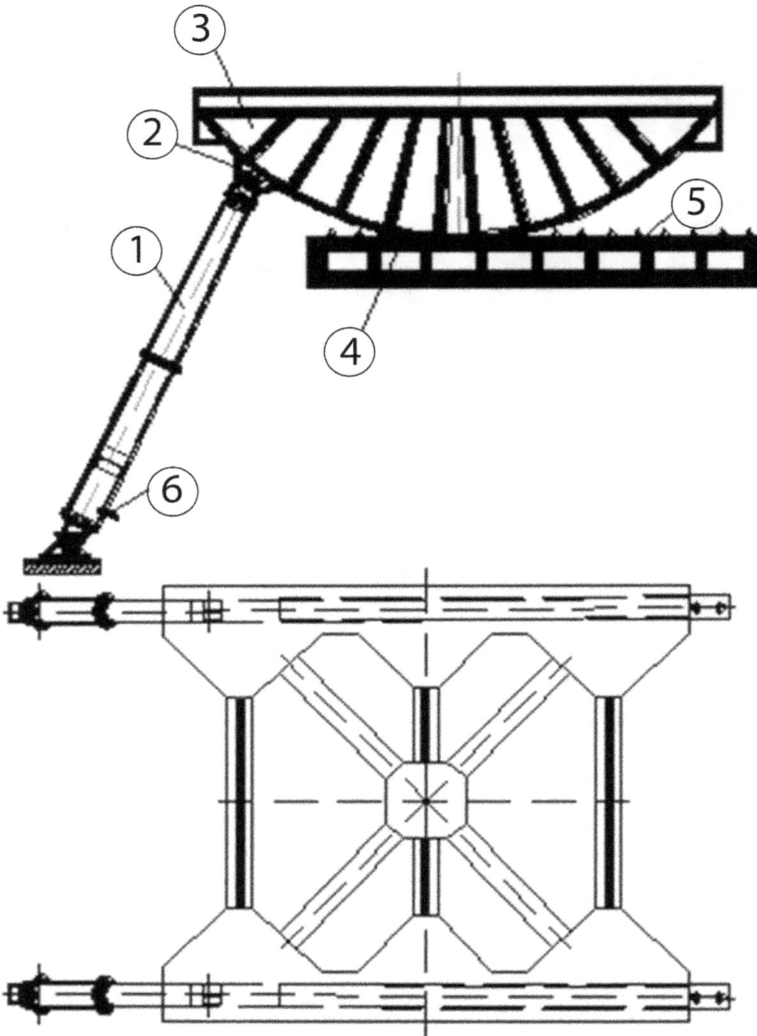

FIGURE 3.43 Design of the furnace tilting mechanism rocker with a hydraulic drive: 1 – hydraulic cylinder, 2 – attachment of the cylinder to the rocker, 3 – rocker, 4 – rocker rail, 5 – conical or pin engaging racks to protect against the rocker slide against the rail, 6 – oil supply to the cylinder. (Author's work.)

electrode lifting mechanisms, and potentially furnace rotation mechanism. The furnace vessel is permanently mounted in a horizontal rocker frame. An electromagnetic stirrer can be installed under the vessel between the rockers. The so-called main platform is made around the furnace. It enables access to the equipment and instruments for furnace operation at the height of the main door.

The roof suspended above the furnace on a support frame on a vertical mast is movable. The mast has the form of a vertical hydraulic cylinder. Therefore, the roof

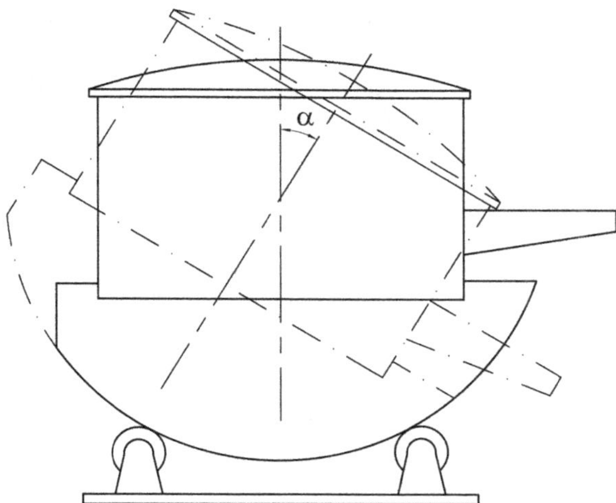

FIGURE 3.44 Design diagram of the roller-type furnace tilting mechanism: continuous line – the furnace in the operating position, broken line – the furnace tilted by an angle α. (Author's work.)

can be raised and lowered. The other cylinder, horizontal, is used for driving its swinging mechanism. Both the roof carrying mast and its swinging mechanism are secured to the rocker structure.

3.6.1 Electrode Holding Equipment

In EAFs, electrodes are suspended vertically in the working space of vessel, so that they can be raised and lowered. Each electrode has its own suspension, which comprises an electrode holder, a horizontal arm, and a mast stem.

The purpose of the holders is to fix the electrodes and to supply electrical current to them. The basic components of the load-carrying structure of the holder are made of a nonmagnetic steel, and the parts for power supply are made of copper or bronze. The friction coefficient between metal and a graphite electrode is low – only 0.15. Therefore, the clamping force must be six to seven times higher than the electrode weight. Clamping shoes are usually made of bronze due to the fact that the electric resistivity of contact between a graphite electrode and bronze is the lowest among other materials. These holders work in difficult thermal conditions as they are heated with a stream of gases escaping around a sealing ring, and a stream of heat flowing from the furnace through the electrode, and eddy currents induced in the electrode. To avoid excessive thermal expansion causing the weakening of the electrode clamping force, the holders are intensively cooled with water. In terms of design, the holders are either jaw type or annular, with a bolted or pneumatic face clamp.

A jaw-type holder consists of a fixed jaw mounted in an arm, to which two clamping jaws are connected in articulation. Both the movable jaw and the fixed jaw are cooled with water. Contact pads are secured to the inner side of the jaws. The power is supplied to the contact pads with a flexible bus consisting of thin copper plates.

It enables the jaws to be opened when the clamping bolt is loosened. The clamping bolt is screwed into nuts placed in clamping jaw pockets. The nuts are fixed in the pockets with adjustable bolts. A design solution like this allows the clamping bolt to be easily replaced in case of its failure or thread damage.

In case of difficulties with aligning the roof to the furnace vessel or if the electrode holes in the roof were shifted during roof laying, the electrode positions can be adjusted with special aligning screws.

Annular holders with face clamping are the most often used design of electrode holders. A holder like this consists of a water-cooled ring, which can be either movable or fixed to the arm of a carrying device, and a movable clamping head built into this ring, or a head fixed to the carrying device arm. The clamping head is also water cooled. The power is supplied to the electrode with four contact pads. Two front pads are secured to the rings, and two rear pads to the movable clamping head. When the clamping head moves forward, the contact pads clamp the electrode. The head clamping force is usually obtained with spring-hydraulic devices.

The electrode holder is secured to the horizontal electrode carrying arm. For larger size furnaces, the arm consists of two parts. The part to which the holder is fixed is water cooled. The arm is constructed in the form of a thick-walled steel tube reinforced with fitted ribs, or it is a hollow beam welded from steel plates and sections. A hollow beam arm is much lighter than a tubular arm with the same load-bearing capacity. The steel for the structure should be nonmagnetic in order to restrict heating with eddy currents from the flowing high intensity current. The arm is also a load-bearing structure for the clamping device and the high current circuit supplying the power to the electrode holder. It is important to electrically isolate the steel arm from the current-carrying conductor.

Often, the so-called conducting arms are used as the electrode carrying arrangement. In a structure like this, load-carrying arms have a form of a thick-walled tube or a hollow beam, and they are made of copper or aluminum. They are water cooled inside. The arm end has a shape adjusted to the electrode camber (circular section with a radius equal to the electrode radius). A movable clamping ring is secured to the arm end. The hydraulic drive of this clamping ring clamps the electrode to the arm end, and the clamping force created in this way is the electrode carrying force. Figure 3.45 shows an example of a design solution of an electrode clamping ring. Then, the electric current flowing to the electrodes goes directly through the load-bearing arm and there is no need to install an additional electric conductor (tubular buses), as is the case for classic load-bearing arms. Therefore, the entire structure of the arm becomes lighter, it requires smaller stands, and cylinders with lower parameters. This solution has a disadvantage, which is that it requires the electrical isolation of the whole arm from the mast structure. Therefore, the mast must be telescopic, and its top end must be flat. A layer of insulating material is placed on this surface, and the arm is placed on this layer and screwed to the mast. Note that, in this case, the whole load-carrying arm is "live" during furnace operation.

Load-carrying devices should enable electrodes to be lifted and lowered at high speeds, which for modern furnaces are at least 6.0 m/min. The distance to which the positioning devices should enable electrodes to be moved depends on the total height of the furnace working space and it is usually from 1.5 to 4.0 m. The structure

FIGURE 3.45 Example of a design solution of an electrode clamping ring for conducting arms [17]. (Advertising materials (corporate brochure) of the SMS group GmbH (formerly Mannesmann Demag Huttentechnik SMS Siemag AG) – with permission.)

of load-carrying devices must be rigid enough and resistant to the rocking of the whole system in the event of its overregulation (e.g. the occurrence of short circuits). The main part of the electrode positioning mechanism are masts, which are usually telescopic, and they are driven by a hydraulic cylinder. The cylinder is usually one-way; it lifts electrodes, which are lowered gravitationally. Oil or water-oil emulsion with a working pressure of approximately 6 MPa is the working fluid. It is supplied from a central hydraulic drive supply station. It is important to use only rigid connections between the cylinder and the workstation. Rapid changes of the working pressure occur in the whole system, and the system operation must be highly reliable. During charge melting, short circuits between the electrode and the charge can occur. A prolonged short circuit can damage the electric power supply circuit. Therefore, the system must guarantee a fast and reliable lifting, which involves high-pressure fluctuations.

3.6.2 Furnace Control Cabin

Control of the individual furnace mechanisms and its electric equipment is necessary for the correct execution of the steel making process, and it is performed from a control cabin. The cabin is located in a place convenient for the furnace operators, in the immediate vicinity of the furnace, so that it is easy to observe the furnace with its auxiliary equipment and to have fast access, especially to the area of the furnace main door.

Basic control accessories are installed in a separate control cabinet, situated on the side wall of the cabin, where the push buttons of the main furnace circuit-breaker, tap-changer, switches of auxiliary mechanism, power input control system switch, and instrumentation are located. A separate cabinet contains protection and signal instruments of the furnace transformer, relays and contactors of the auxiliary mechanism drive power supply system, electric arc automatic power control relays, and indicating and recording instruments for current and voltage, active and reactive power, electricity consumption, etc.

A separate control cabinet contains electrode lifting and lowering limiters, limiters of the roof lifting during its removing from the furnace vessel, furnace tilting mechanism, etc. The control board of this cabinet contains the necessary measurement instruments for the control of the thermal operation of the furnace and automatic control of the whole steelmaking manufacturing process. In addition, for some mechanisms, the possibility of switching from automatic to manual control is provided in case of a failure.

The control panel is the most essential equipment of the control cabin, and it is equipped with a monitor for the visualization of a diagram showing the operation system of key mechanical, electrical components, and the manufacturing process. The operator is responsible for the correct execution of the manufacturing process, failure-free operation of all the equipment, and safe working conditions, and they must be able to carry out continuous observation and monitoring of all the parameters of the steel melting process in the EAF. Therefore, all of the most important equipment operation parameters and the process technological data are displayed on the monitor-type panel. When any acceptable values for any parameter are exceeded, it is signaled in a clearly visible way, usually also audibly, to quickly draw the operator's attention. The operator has a control keyboard, which allows them to enter, set, or correct the technical and technological parameters of the heat.

REFERENCES

1. Advertising materials (corporate brochure) of the SMS group GmbH (formerly SMS Siemag AG).
2. Karbowniczek M.: Electrometallurgy of steel; Exercises, course book of AGH University of Science and Technology Krakow, Poland, no. 1364, 1992 (in Polish).
3. Verdeja L., Sancho J., Ballester A., Gonzalez R.: *Refractory and Ceramic Materials*, Editorial Sintesis, Madrid, 2014.
4. Advertising materials (corporate brochure) of Didier Company.
5. Advertising materials (corporate brochure) of Radex Company.
6. Advertising materials (corporate brochure) of Zaklady Magnezytowe "Ropczyce" Company.
7. Schwabe W. E.: Arc heating transfer and refractory erosion in the electric steel furnace, *Proceedings of Electric Furnace Conference (AIME)*, Cincinnati, OH, Vol. 20, pp. 195–206, 1962.
8. Jones J., Bowman B., Lefrank, P.: Electric Furnace Steelmaking, in *The Making, Shaping and Treating of Steel*; R.J. Fruehan Editor, The AISE Steel Foundation, Pittsburgh, PA, 1998, pp. 525–660.
9. Frittella P., Colombo V., Ghedini E., Solari G.: Application of 3-D time dependent modelling of plasma arc to an AC Electric Arc Furnace, *Proceedings of the 10th European Electric Steelmaking Conference*, Graz, Austria, 2012.
10. Majewski Z., Pogorzalek J., Zdonek B.: Dotychczasowe doswiadczenia w stosowaniu chlodzenia wodnego scian duzych piecow lukowych, *Hutnik*, Vol. 12, 1983, pp. 451–457 (in Polish).

11. Majewski Z., Zdonek B., Pogorzałek J.: Wodne chlodzenie sklepien piecow elektrycznych lukowych, *Hutnik*, Vol. 1, 1985, pp. 29–33 (in Polish).
12. Paliy G., Zinurov I., Sokolov A.: Wljanie położenija dug na raspredelenie teplowych potokow w elektrodugowoj peci, *Izwestija Wyssich Ucebnych Zawedenij Czernaja Metallurgija*, 1975, Vol. 3, s.85–s.88 (in Russian).
13. Kadinow E., Tudier I.: Mietodika rasceta racionalnich parametrow plawilnogo prostranstwa i wnutrennogo profilja stien dugowyh staleplawilnych piecej, *Metallurija i Koksochimija*, 1975, Vol. 44, s.30–s.35 (in Russian).
14. Spelicin R., Smoljarenko W., Kurlykin W.: Prawomiernost primienienija zakona Keplera dlja rasceta oblucennosti futierowki DSP, *Elektrometallurgija*, 1976, Vol. 6, s.6–s.7 (in Russian).
15. Hering M.: *Termokinetyka dla elektrykow*, Wydawnictwa Naukowo – Techniczne, Warszawa, 1980 (in Polish).
16. Pawlik T., Slomska I.: Technika cieplna, AGH course book no. 583, Krakow, 1977 (in Polish).
17. Publicity materials of Mannesmann Demag Huttentechnik.

4 Electric Equipment of EAFs

Depending on the practice applied, the electricity consumption during steelmaking in electric arc furnaces (EAFs) ranges from 300 to 700 kWh/t, which corresponds to approximately 1,100–2,500 MJ/t of steel produced. In the electric arc, this energy is transformed into heat energy which is used for running the metallurgical process and for making up for the heat losses accompanying this process. To ensure stable operating conditions, EAFs are equipped with adequately adapted power supply systems and electric arc operation control systems. At the same time, these systems allow the metallurgical process of steelmaking to be continuously monitored. An electrical steel shop and its installed furnaces are supplied directly from a high-power grid. This solution guarantees, on one hand, a constant supply of electricity to the steelworks, and on the other hand, predictable demand for the electricity producers. It creates convenient conditions for the systematic and planned utilization of installed furnace capacities [1].

4.1 POWER SUPPLY SYSTEM

In order to minimize excessive electrical energy losses during power transmission from a power station to an electric steelmaking plant, medium voltage (e.g. 15 or 30 kV), high voltage (e.g. 110 kV), or ultra-high voltage (e.g. 220 or 400 kV) power transmission lines are applied. The type of power line supplying an EAF is selected depending on the furnace size and the related power demand as well as the distance to the so-called transmission substation. Small furnaces, equipped with low-capacity transformers, are supplied from grids with voltage up to 30 kV, whereas large furnaces with higher capacity transformers from grids with voltage of at least 60 kV.

Depending on the power supply grid voltage, the power supply system of an EAF consists of either one part or two parts. For voltages above 30 kV, a two-part system is applied, including the transformation from a high voltage to an intermediate voltage (usually 30 kV) and the furnace direct power supply system. For voltages under 30 kV, only the furnace direct power supply system is applied (Figure 4.1).

From the point of view of furnace operation, only the direct power supply system is relevant. The first device in this system is a disconnect switch and the next is a furnace operating switch, which is followed by the most important component – the furnace transformer, which transforms the grid voltage or intermediate voltage to a low voltage, to which the so-called secondary circuit is directly connected. The secondary circuit consists of bus bars, flexible cables, and tubular buses. The tubular buses are connected directly to graphite electrodes, with electric arcs burning at their tips. The system includes measurement instruments, monitoring and control instruments, and assemblies of devices for reactive power compensation and filtration, as well as

DOI: 10.1201/9781003130949-4

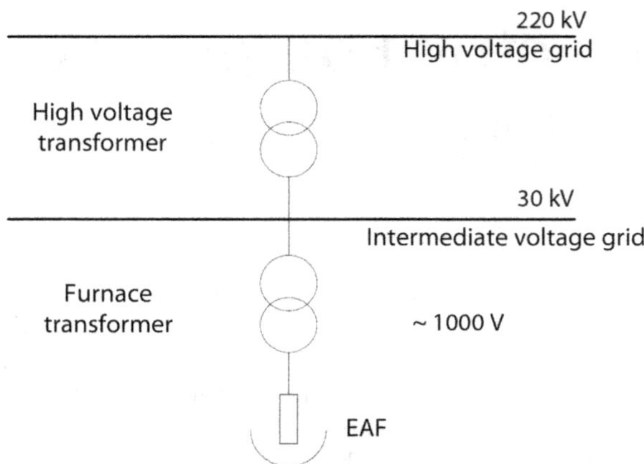

FIGURE 4.1 Schematic diagram of the electric power supply of an electric arc furnace. (Author's work.)

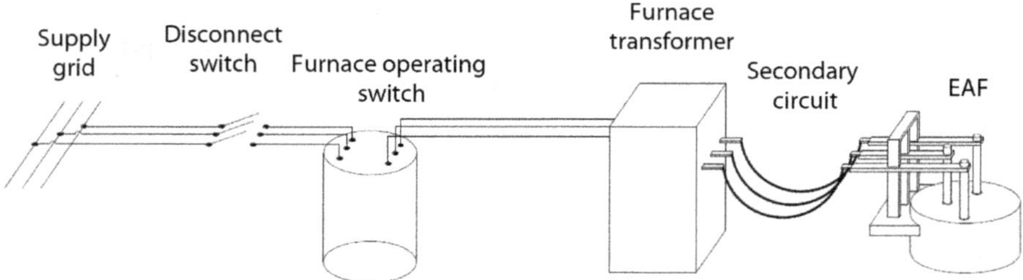

FIGURE 4.2 Simplified diagram of the electric arc furnace power supply system. (Author's work.)

stabilization of the electric arc. The whole system can be divided into the primary circuit (upstream the furnace transformer) and the secondary circuit (downstream the furnace transformer). A simplified general diagram of the EAF power supply system is shown in Figure 4.2.

A connection diagram of the furnace power supply system, starting with high voltage, including measurement systems is shown in Figure 4.3.

Characteristic current parameters of the power supply system, such as voltage, current intensity, active power and reactive power, and energy consumption, are measured on the primary and secondary side of the furnace transformer. All measuring instruments are connected to the circuit by a relevant voltage transformer on the primary side, and a current transformer on the secondary side. The measured values are transmitted to the control and monitoring systems as well as to the operator's panel, where they are used to monitor the correctness of the manufacturing process. At the same time, the measured values constitute fundamental input data for control systems of the whole system-operating parameters, protections against system damage, and data collection and recording for settlements with the power supplier, cost analysis, etc.

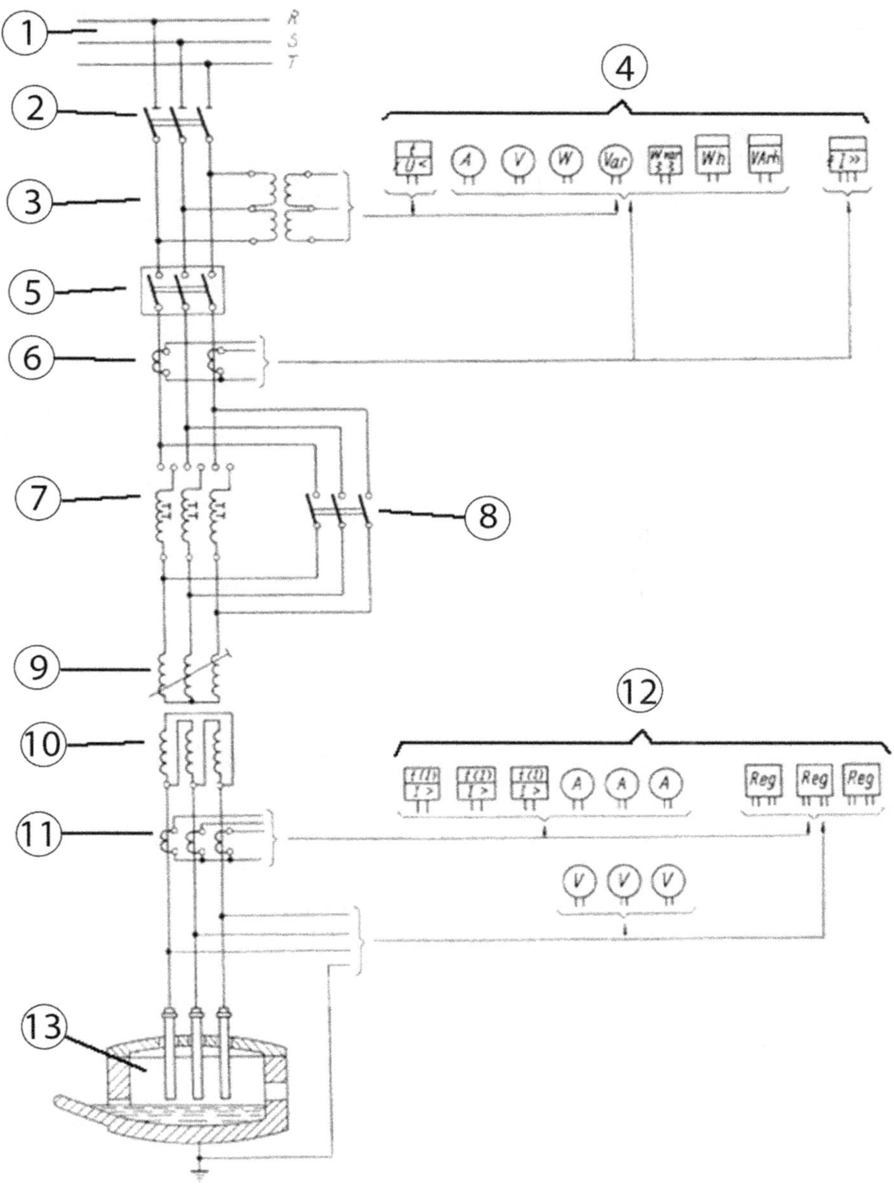

FIGURE 4.3 General connection diagram of electric arc furnace power supply system: 1 – three-phase power grid, 2 – disconnect switch, 3 – voltage transformer, 4– measuring instruments on the primary side, 5 – furnace operating switch, 6 – current transformer, 7 – reactor, 8 – reactor switch, 9 – furnace transformer primary winding, 10 – furnace transformer secondary winding, 11 – current transformer, 12 – measuring instruments on the secondary side, 13 – electric arc furnace. (Author's work.)

4.2 DISCONNECT SWITCH AND FURNACE OPERATING SWITCH

The furnace disconnect switch is predominantly made as an overhead switch, and it is located in a field substation outside the steelmaking shop building. Overhead wires from the power line are connected to the disconnect switch. Therefore, the voltage from the power grid to the furnace power system can be on or off. The disconnect switch complements the operation of the furnace operating switch. It is used for closing and opening electric circuits with no current. The disconnect switch disconnects or connects the voltage to the system, and these operations can only be practically performed at zero current. The furnace operating switch is for switching an energized (on load) furnace on or off. The purpose of the disconnect switch is to "cut off" the voltage upstream the furnace operating switch. It is seldom used, only during maintenance or service inspections on the furnace operating switch as well as in emergency situations. Often, for safety reasons, gear drives are interlocked so that the disconnect switch cannot be disconnected when the furnace operation switch is on.

The design of a disconnect switch is much simpler than that of an operating switch, as it primarily performs the role of an isolating switch. For an EAF, disconnect switches are mounted on the premises of the so-called electrical substation, outside the steelmaking shop building. As they operate outdoors, they are called overhead disconnect switches. Their construction includes a structural base placed on a foundation, stand-off insulators, movable and fixed contacts, and drive components for driving movable contacts. The design of a disconnect switch should ensure a visible circuit break between its contacts. The contacts are made from copper, sometimes silver-coated, blades or clamps, with an adequate clamping force to ensure electrical connection with a minimum resistance. The main operating parameters of disconnect switches applied in EAF power systems comprise rated voltage and operating current intensity, adjusted to the transformer capacity. In terms of the contact closing and opening method, there are blade contact, rotating-type, pantograph, articulated, or slide switches. In terms of positioning, there are horizontal or vertical switches [2].

A furnace operating switch, which can disconnect an electric circuit during the current flow, consists of connecting components such as copper blades and clamps, and a system for quenching the arc forming during the circuit breaking. The whole is enclosed in a housing of special design. There are various designs and structures of furnace operating switches, depending on rated parameters, including voltage and rated current, short-circuit current intensity, and the allowed operation temperature. For EAFs, medium- and high-voltage switches are applied. The switches can be classified by a medium used for electric arc quenching. Here, we use the following designs of high-power switches, used in arc furnace electric circuits [3]:

- oil (older designs),
- air,
- vacuum,
- sulfur hexafluoride (SF_6) single-pressure (new solutions),
- sulfur hexafluoride (SF_6) self-blasting (state-of-the-art solutions).

The furnace operating switch works in particularly difficult conditions during the charge melting period. At the beginning of the melting period, when the charge is still cold, and "fairly loose", the operation of electric arcs is rather unstable. Later, as a result of the charge cave-in, very often short circuits and current "hits" occur, causing short-term overloads of the transformer. To avoid the premature wear of the transformer, an oil circuit breaker automatically trips the furnace when the current exceeds the limit value. For this purpose, the power supply system has an embedded current regulator, whose work is adjusted to a specific time of current overload and its value. As the current increases, the acceptable transformer overload time decreases.

Usually, a power cable is used to make the electric circuit from the operating switch to the transformer. The cable is placed in a dedicated duct, usually made in the floor of the room in which both the switch and the transformer are located.

4.3 FURNACE TRANSFORMER

4.3.1 The Principles of Alternating Current Transformation

The transformer is built of a magnetic core and primary and secondary winding coils, which are wound on the core. Alternating voltage connected to the primary winding produces an electromagnetic field inside the coil and around it. The field mainly covers the core due to its high magnetic conductivity. Some electromagnetic field energy is used to cover the loss related to the primary winding resistance and reactance, and the compensation of the "reverse" electromotive force induced in this winding. At the same time, the electromagnetic field within the core induces the electromotive force in the secondary winding of the transformer, and this force generates voltage on the coil terminals. Some electromotive force is used to cover the loss related to resistance and reactance of the secondary winding. We can make a simplified assumption that the relationship between the primary and secondary winding voltages is expressed in the following formula:

$$\frac{U_p}{U_s} = \frac{N_p}{N_s} \quad \text{or} \quad U_s = \frac{N_s}{N_p} \tag{4.1}$$

where:

U_p and U_s – primary and secondary winding voltages, respectively, V
N_p and N_s – number of turns in the primary and secondary windings, respectively

As can be seen in the above-mentioned relationship, the value of voltage in the secondary winding coil is proportional to the voltage in the primary winding and the ratio of turns in the secondary and primary coils. A similar dependence on the ratio of turns applies to the current intensities in both the windings:

$$\frac{I_p}{I_s} = \frac{N_s}{N_p} \quad \text{or} \quad I_s = \frac{N_p}{N_s} \tag{4.2}$$

where:

I_p and I_s – current intensity in the primary and secondary windings, respectively, A

This means that the current intensity in the secondary winding is proportional to the current intensity in the primary coil and the ratio of turns in the primary and secondary winding coils. From the above, rather simplified analysis, it follows that the transformer enables the energy to be "transferred" from the primary side to the secondary side while, at the same time, changing the parameters of this energy and with certain losses. Energy losses are related to magnetic and electric losses. The magnetic losses stem from the flow of electromagnetic flux through the transformer core, while the electric ones stem from the current flow through the coil windings. The total loss does not exceed a few percent of the energy transformed.

4.3.2 Design of Furnace Transformers

EAFs are supplied from three-phase alternating current transformers. These transformers operate in much harsher conditions than the usual transformers due to the occurrence of frequent short circuits in the furnace electric circuits. These short circuits occur not only during the ignition of electric arcs but also during melting as a result of shifts of scrap that has melted. During short circuits, considerable mechanical forces occur in the transformer winding and its housing. Therefore, this winding should be sufficiently strongly fixed on the magnetic core, and the core itself should be fixed in the transformer housing so that its movement in any direction is impossible. Furnace transformers, as the most important components of the power supply system, should have the following features:

- large transformation ratio, i.e. the ability to step down from a very high voltage to a low voltage (max. 1,200 V) and a high current intensity,
- control of the power input that is as smooth as possible, adjusted to the current needs of the metallurgical process in progress,
- ability of a long-term overload (up to 125% power rating), without the fear of damaging the winding insulation as a result of an increase in the temperature,
- high mechanical stability in relation to variable dynamic stress resulting from short circuits in the furnace power supply system.

Furnace transformers, in particular of high power, are custom designed and made for each steelmaking shop. A furnace transformer comprises an oil filled tank, in which the core, on which the primary and secondary winding coils are wound, is immersed. The tank is made of steel plate in the form of a container with radiator fins increasing the cooling surface.

Isoparaffin mineral oils are used for cooling the transformer windings, whereas in newer solutions either synthetic oils or natural liquid esters (less harmful to the environment in case of a leak) are applied. The oil circulation is forced – oil heated in the transformer tank is pumped to a coil pipe in a water cooler located outside the tank housing. The pressure of oil in the transformer and the cooler is higher than the pressure of the cooling water. It prevents the penetration of the cooling water to the oil and the deterioration of its insulating properties. The oil temperature in the transformer must not exceed 348–353 K (75–80°C). When this temperature is exceeded, the transformer winding insulation life substantially shortens. An oil reservoir is located above the transformer, and oil in the transformer housing is replenished from this reservoir. At the same time, it limits the area of direct contact of oil and air, and thereby its oxidation. Therefore, the useful life of the oil is increased. Before the oil is poured into the tank, it needs to be treated by thorough drying, degassing, and removing fine solid impurities.

The cores are made of silicon steel sheets (so-called transformer steels) with a thickness of approximately 0.3 mm. Steels like this feature a fine, oriented grain structure that ensures a good conductivity for the magnetic flux. Individual thin steel sheets, which form the core structure, are electrically isolated from one another. Sheets for making the core are cut mechanically and then manually assembled to ensure the required quality of their correct arrangement.

The coil windings are made of copper wires. Windings are wound on vertical or horizontal winding machines, with a very precise turn shaping and very accurate tight fit. Sometimes, an axial pressure is also applied during winding. The coils are also dried, often with additional oil proofing. The ends of windings are taken outside of the tank and isolated electrically with the so-called bushings. Bushings are soldered to the coils and then ended with special terminals enabling power cable wires on the primary side, or secondary circuit bus bars to be connected.

In addition, transformers are fitted with on-load tap changers, current transformers, so-called oil conservators and temperature indicators. On-load tap changers are often installed inside a transformer in a hermetic enclosure in the form of a tight cylinder of fiberglass-reinforced epoxy resin. The tap changer oil has no contact with the transformer oil. New solutions contain vacuum extinguishing systems, which extends the tap changer life. Tap changers with contacts working in oil are often fitted with filtering systems. Desulfurized and dehydrated oil increases the contacts life.

Core transformers with vertical columns are the most common design. They have cylindrical windings. Radial electrical forces, occurring during short circuits, cause tension in the external turns and compression of internal turns. Therefore, an adequate mechanical strength of the windings is necessary. The suitable dielectric strength of insulation is very important. The furnace transformer is continuously exposed to long-term and variable intensities of electric fields. Windings are insulated with soft cellulose saturated with oil, hard paper (insulating board), and saturation oil. Before installation, cellulose-insulating materials must be well dried and pressed to ensure their adequate mechanical strength.

Transformer operating parameters largely depend on the method of winding preparation and their geometry as they are decisive to the value of inductive reactance of the entire furnace supply system. Inductive reactance is an important parameter of the power supply system. The higher the inductive reactance, the better the conditions for stable arc burning; however, the energy losses are higher. For each power supply system, depending on the operation conditions of a specific furnace, there is an advantageous value of reactance. Therefore, the transformer winding is designed and made individually for each furnace so that its reactance is optimal for its operating conditions. Usually, the coil windings of high-power transformers, for large furnaces, have sufficient inductance. However, in some cases, in particular for low-power transformers, which are designed for supplying low-capacity furnaces, the coil windings have insufficient reactance and, therefore, additional induction coils (reactors) are embedded in their circuits. Transformers with embedded reactors have two motor drives: one for the tap changer used for the voltage control, and the other one to disconnect the reactor. Transformers that work with separate reactors have one motor drive for a tap changer, which is used for the voltage control. Motor drives are designed for remote control [4].

In order to detect any damage of the transformer winding insulation at an early stage, special gas sensors are installed. As a result of insulation damage (its ignition) or thermal decomposition of oil, gases appear in the transformer. Then, the gas sensors react by automatically transmitting a relevant acoustic or visual signal to the furnace control cabin. Similarly, when the oil level falls, an acoustic or visual signal is generated. Intensive gas emission causes the generation of an additional pulse, which is used for tripping the main oil circuit breaker and thereby to disconnect the furnace from the power grid.

There are various methods of making the core and connections of furnace transformer windings. A system of two magnetic cores in a single housing is one of the solutions, especially for small capacities. On one core, there is a furnace transformer winding with a fixed transformation ratio and, on the other one, there is a control autotransformer winding. Where these windings are installed in separate housings, secondary connections (bushings) are simplified, which facilitates the simple and symmetrical execution of the furnace secondary circuit. However, the solution with a separate furnace transformer and a separate control transformer is more expensive. The assembly of a fixed transformation ratio furnace transformer and a control transformer allows the power input to be controlled at the same voltage step, and an auxiliary winding for connecting a capacitor battery to improve the power factor ($cos\varphi$) to be installed. The tap changer should consist of three separate tap changers, individually for each phase. In this version of transformer winding, the short-circuit voltage strongly increases when the secondary voltage decreases, and reactance X_s remains practically constant:

$$U_z = \frac{I \cdot U_{s\max}}{U_s} \tag{4.3}$$

In addition, a solution with a main transformer and a secondary transformer connected in series is applied. Then, the short-circuit voltage U_z=constant, whereas: $X_s = \frac{U_s}{U_{s\max}}$

where:
U_z – secondary side short-circuit voltage, V,
U_s – secondary side voltage, V,
$U_{s\max}$ – maximum value of the secondary side voltage, V,
I – current intensity on the secondary side, A,
X_s – secondary side reactance, Ω.

A system like this can be directly connected to the high-voltage grid (110 or 220 kV) and also provides equal voltage taps. Connections between the main and secondary transformers increase electrical losses and costs compared to other solutions, in particular when secondary currents achieve high values. In certain design solutions, these connections can be useful.

Furnace transformers designed for the direct control of power input are also applied. Then, the low-voltage (secondary) winding is situated between the high-voltage (primary) winding and the control winding. In this system, a change in reactance X_s is roughly proportional to the factor $y = \frac{U_z}{U_{s\max}}$, and the short-circuit voltage remains practically constant in the whole range of power input control. Due to the method of manufacturing, the transformer secondary winding, the current intensity in these solutions cannot exceed 20 kA. Transformers with the secondary winding outside the control winding do not have such a limitation. In this solution, reactance X_s at all taps is lower than for $U_{s\max}$.

The stability of arcing depends on the value of the furnace system reactance X, which consists of the reactance of the secondary circuit connections and of the furnace transformer. Reactance in low-capacity furnaces is insufficient to ensure stable electric arc burning when the charge is cold, and to mitigate the effects of short circuits occurring in the initial period of scrap melting. To stabilize electric arc operating conditions, a reactor is installed in the furnace system upstream of the transformer. It has specific values of resistance and reactance. The reactor comprises a magnetic core with an induction coil wound on it. They are placed in a tank filled with oil. The reactor reactance is usually from 25 to 30 times impedance Z of the furnace power supply system, corresponding to the rated current intensity. The reactor coils have outside tips, thereby allowing its reactance to be adjusted every 5%, 10%, 15%, 20%, and 25% of impedance Z of the furnace power supply system.

Operation of the furnace with the reactor on is disadvantageous in terms of energy, as it causes a reduction of the power factor value. Therefore, the reactor is shunted (excluded from the circuit) once the arc operation becomes stable, which usually happens after about 30% of the melting time. The reactors are installed at furnaces supplied from transformers with a capacity under 10 MVA. They can constitute a separate structure or are built in directly in the furnace transformer.

To obtain more voltage steps at the secondary side of the transformer, and thereby to be able to control the input power in a broad range of its values, the primary winding coils are made in segments, and their tips constitute the so-called transformer taps. The primary winding coils can be periodically reconnected from delta to star and vice versa.

The transformer secondary voltage changes inversely proportional to the number of turns in the primary winding. To increase the secondary voltage, the number of turns in the primary winding should be reduced accordingly. The reduction of the turn number causes an increase in the magnetic induction in the steel core of the transformer. To control the power input, a special design tap changers or autotransformers installed in the transformer housing are applied. Tap changers should enable the power input to be adjusted on load. Autotransformers can be either single phase or three phase. The autotransformer features a connection of primary and secondary windings through not only the magnetic core but also a direct electrical connection. Thereby the secondary winding coil is at the same time a part of the primary winding and vice versa.

4.4 SECONDARY CIRCUIT

The secondary circuit comprises conductors supplying power from the points of the furnace transformer secondary side to the contact pads of graphite electrode holders. It consists of:

- bus bars,
- flexible cables, and
- tubular buses.

Each of these sections performs different functions and works in different conditions. The circuit arrangement like this, apart from conducting the current, enables the fixed transformer to work with the electrodes moving in both the horizontal and vertical planes. A general layout of the secondary circuit is shown in Figure 4.4.

Bus bars are used for supplying power from the points (the ends of windings) of the furnace transformer secondary side to the flexible cables. They comprise copper bars in the shape of uniform flat bars or sheets laid in packages, being also in the shape of a flat bar. Due to energy losses, they should be as short as possible.

Flexible cables supply power from the bus bars to the tubular buses. The cables must be suitably flexible to enable electrodes to freely move in the vertical direction and to swing or shift together with the furnace roof in the horizontal plane. This section of the circuit has a "U" shape and is made of a special, tubular structure of flexible cables, cooled with water [5].

From the outside, the cable has a water hose made of synthetic rubber reinforced with layers of nylon. This component has adequate mechanical properties as regards twisting and shear stress occurring in the cable during furnace movements and the resistance to internal and external mechanical shocks of electrodynamic origin from a fast changing electromagnetic field. It enables water to flow at a pressure from

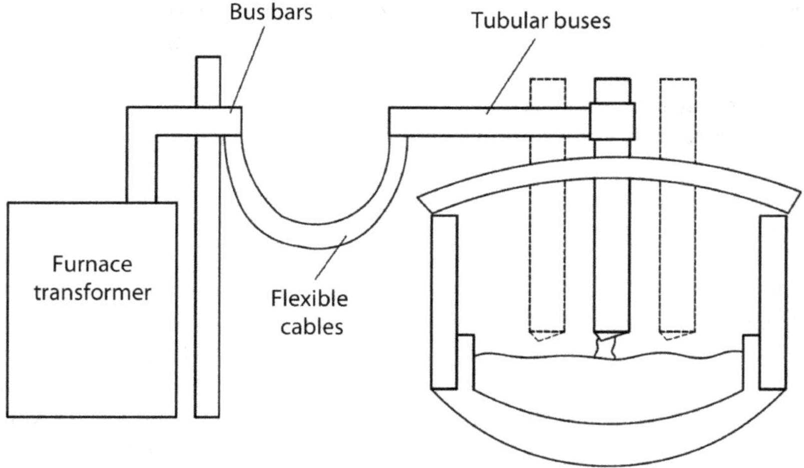

FIGURE 4.4 General diagram of the secondary circuit. (Author's work.)

0.4 to 0.6 MPa. On the outside, the hose can be covered with anti-chafing and anti-radiation materials. Insulating properties of the hose enable a voltage up to 1,200 V to be applied.

There is an electric conductor inside the hose that is made of electrolytic (so-called oxygen-free) copper, ensuring adequate electric conductivity [5]. The conductor consists of a set of a dozen or so lines that are made of single wires. The lines are arranged concentrically, often in a few rows on a core made of synthetic rubber. The rubber core is a load-carrying component for the copper conductor, ensuring its round shape. There are steel springs inside the core to ensure a connection between the rubber core and copper tips (terminals) and to damp vibrations in the tip zone.

Flexible cables are ended with special copper tabs (terminals) that enable other circuit elements to be connected. The cables are individually designed and manufactured to match the design of each furnace. It is particularly important to fit the length of the flexible cable to a specific furnace. Conductivities of the flexible cables range from 3 to 20 $\mu\Omega$/m.

Another design of the flexible cable is the case where the rubber pipe is outside the cable, and this tube insulates and shields the copper conductors [5]. An internal pipe, also made of rubber, provides the flow of the cooling water. The design of copper conductors is similar to the former case.

Tubular buses are used for supplying electric current from flexible cables to the contact pads of electrode holders. They are made of copper tubes cooled with water and are installed on the arm of the electrode-carrying device. One end of tubular buses is adapted to connect the tab of flexible cables. The other end of tubular buses is fitted with an electrode holder with contact pads covered with a copper foil package. The holder with pads must ensure a good electric contact of the bus with the electrodes in order to minimize the electric losses of a high current flow. The graphite electrodes are carried by load-bearing arms made of a steel structure and ended with electrode holders. The holders are most often designed in the form of a movable, hydraulically driven steel clamping ring. The ring clamps the electrode to the fixed part of the load-carrying arm, and the friction force between the holder and the electrode ensures the adequate load-carrying force. A view of tubular buses is shown in Figure 4.5.

At the bottom, you can see the load-carrying arms made of steel structures, to which copper tubular buses are secured. In the front, you can see the contact pads for securing graphite electrodes along with electrode holders.

Modern solutions introduce a new design, in which tubular buses are replaced with the so-called conductive load-carrying arms. Such arms conduct the current and carry the electrodes at the same time. Arms like these have a shape of a "rectangular pipe", and they are made of copper or aluminum [7]. Regardless of the solution, the secondary circuit is ended with contact pads clamped to the graphite electrodes, enabling the current to be supplied to them with a low contact resistance.

The design of the secondary circuit should minimize the electrical energy losses occurring in the circuit (high electric efficiency) and should ensure the uniformity of power distribution between the individual phases (low power asymmetry). To meet these requirements, the secondary circuit should feature a low value of resistance and reactance, and the inductance of its individual phases should be equal. At the same time,

FIGURE 4.5 View of tubular buses arranged on load-bearing arms [6]. (Source: Advertising materials (corporate brochure) of the SMS group GmbH (formerly SMS Siemag AG and Mannesmann Demag) – with permission.)

to obtain stable conditions of electric arc generation, the secondary circuit inductance should not be lower than the characteristic value of a furnace with a specific capacity and power. Usually, larger furnaces have a sufficient inductance. In smaller furnaces, a reactor is connected in series to the transformer primary winding circuit to increase the furnace inductance during the charge melting (unstable operation of electric arcs). The design and work quality of the secondary circuit should minimize the consumption of materials while maintaining the circuit reliability and ease of operation.

In industrial practice, various design solutions of the secondary circuit arrangement are employed. The main purpose of the optimal structure is to ensure the so-called symmetry in terms of electric impedance of individual conductors. The electric impedance of a conductor involves the vector sum of resistance and reactance. The resistance is related to material properties of the conductor (resistivity) and its geometric dimensions. For the secondary circuit, the length of individual conductors is important. In the EAF, the electrodes are located symmetrically on the circle perimeter, inside the furnace; therefore, the distance of the individual electrodes from the transformer is not equal, thereby causing different lengths of tubular buses. This results in different resistances of individual tubular buses, and it is not possible to design a fully symmetrical system (Figure 4.6a and b). Either a system with one long conductor and two shorter ones (Figure 4.6a) or a system with one shorter and two longer conductors (Figure 4.6b) is possible.

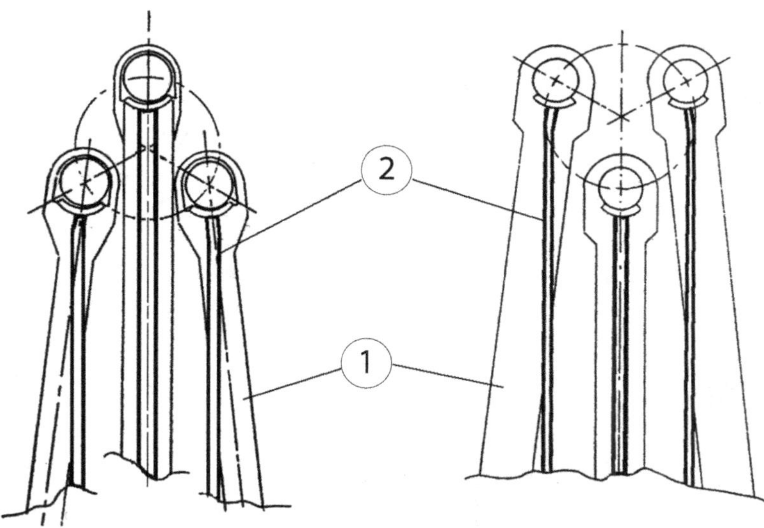

FIGURE 4.6 Diagram of a secondary circuit conductor layout – plan view [8]. (a) System with one longer and two shorter conductors. (b) System with one shorter and two longer conductors: 1 – carrying arm, 2 – conductor. (Adopted from Charles R. Taylor (editor), 1985 – with permission.)

Reactance in the case of EAF primarily consists of inductive reactance, depending on self- and mutual inductance of conductors leading the electric current. There are two basic design systems of the secondary circuit: coplanar and triangular. A diagram of these systems is shown in Figure 4.7. The inductance of each conductor is the resultant of self- and mutual inductance. Self-inductance of all conductors is approximately the same, whereas the mutual inductance depends on the distances between the conductors leading the current. Simplified, we can assume that for the coplanar array inductances in individual conductors of the secondary circuit are as follows:

$$\begin{aligned} L_A &= L_{\text{self}} - L_{\text{mutual Bm}} - L_{\text{mutual Cm}} \\ L_B &= L_{\text{self}} - L_{\text{mutual Am}} - L_{\text{mutual Cm}} \\ L_C &= L_{\text{self}} - L_{\text{mutual Am}} - L_{\text{mutual Bm}} \end{aligned} \tag{4.4}$$

where $L_{\text{mutual Bm}}$ (and the other ones analogous) means the mutual inductance between conductor A or C relative to conductor B, where the distance between the conductors is m. It can be seen from the above that $L_A = L_C$ and $L_B < L_A$ and $L_B < L_C$, which arises from unequal distances of individual conductors. Therefore, the reactances of individual conductors in a system like this are not equal. However, for a triangular system, all of the distances between the conductors are equal; thus, both self and mutual inductances are equal and the resulting reactances of the individual conductors are the same.

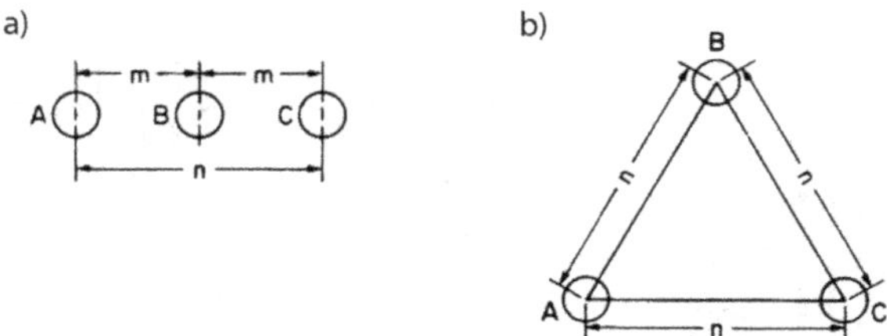

FIGURE 4.7 Diagrams of a secondary circuit conductor layout in a cross section [8]. (a) Coplanar. (b) Triangular. (Charles R. Taylor (editor), 1985 – with permission.)

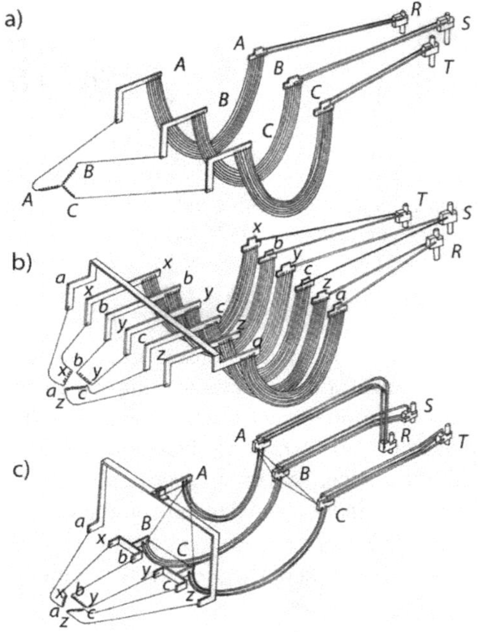

FIGURE 4.8 Diagrams of various design solutions of the secondary circuit. (Author's work.)

Various examples of potential design solutions of the secondary circuit are shown in Figure 4.8a–c. On the left-hand side of all the drawings, furnace transformer secondary winding coils are shown schematically (Figure 4.8a) in the star connection, whereas in Figure 4.8b and c in the delta connection. Next, various possible designs of bus bars, flexible cables, and tubular buses are shown. Each of the demonstrated solutions has its advantages and disadvantages. In industrial practice, the circuit design solution is adapted individually to the specific furnace operating conditions, in particular the construction of the furnace, arrangement of the equipment in the steelmaking shop, melting practices applied, and the range of steels manufactured [9].

4.5 INSTRUMENT TRANSFORMERS

Instrument transformers enable measuring and protecting instruments to be connected to a high-voltage or high-amperage grid. They allow these instruments to be safely operated and installed at a remote distance from the location where a specific parameter is measured, and thus to be placed on a panel in the control cabin for the operators. The application of instrument transformers allows us to use a relatively small number of standard types and varieties of measuring instruments. The design of an ammeter for measuring current with an intensity of a few thousand amperes is technically difficult. Thanks to an instrument transformer, this measurement can be made with an ammeter for measuring current with an intensity up to 5 A. The case is similar when voltage is measured. Instrument transformers, due to differences in their construction, are divided into voltage and current transformers.

The measuring instruments or selected coils of the measuring instruments whose readings should be proportional to voltage, i.e. voltmeters, voltage windings of wattmeters, and voltage relays, are connected to voltage transformers. The primary winding of a voltage transformer consists of a large number of turns and it is connected in parallel to the network in which the measurement is made. The secondary winding consists of a small number of turns. The components of measuring instruments are connected to this winding. The voltage on the secondary side of an instrument transformer is:

$$V_s = V_p \frac{N_s}{N_p} \tag{4.5}$$

where:

V_p, V_s – voltage of the primary and secondary side of a voltage transformer, respectively, V,

N_p, N_s – number of turns of the primary and secondary windings of a voltage transformer, respectively.

Knowing the ratio N_s/N_p, the actual value of the grid voltage V_p can be established on the basis of the reading of value V_s. Very often, the measuring instrument has a scale adapted to read the value V_p directly. A voltage transformer is designed in a way that its reactance is as low as possible, while the resistance of the measuring instruments connected to it should be as high as possible. Thereby, the current flowing in the measuring circuit is low and the measurement error obtained is negligible.

The number of measuring instruments that can be connected to a voltage transformer depends on its power. The total power of the connected measuring instruments cannot be higher than the rated instrument transformer power. As the voltage transformer reactance is low, voltage transformers are very sensitive to short-term short circuits. Therefore, the primary winding coil is connected to the grid through a circuit breaker and an additional system of resistors. The secondary winding coil, for occupational safety reasons, should be grounded in case of its shorting with the primary winding coil, and to avoid hazards related to an occurrence of capacity current between these coils.

These measuring instruments, through which a current proportional to the current intensity in the transformer supply grid flows, are connected to the measuring current transformers. These instruments include ammeters, current intensity windings of wattmeters, current relay windings, etc. The primary winding of this transformer consists of a small number of turns and is connected to the grid in series. The secondary winding consists of a large number of turns and measuring instruments are connected to its circuit in series. Instrument readings are based on the relationship:

$$I_s = I_p \frac{N_p}{N_s} \tag{4.6}$$

where:

I_p, I_s – current intensity of the primary and secondary side of a current transformer, respectively, A,

N_p, N_s – number of turns of the primary and secondary side of a current transformer, respectively.

Measuring instruments are usually calibrated directly in the scale of measured values of the primary current intensity. The more similar to a short circuit in the secondary circuit the transformer operating conditions, the higher the accuracy of its current intensity measurement. In fact, these conditions are satisfied because usually the resistance of ammeters and other metering instruments is low and, therefore, a current transformer practically operates in the short-circuit state.

When selecting a current transformer, the rated voltage and amperage of the secondary circuit, short-circuit current, and the multiplication factor of the secondary short-circuit current are taken into account. The multiplication factor of the secondary short-circuit current is relevant for the operating conditions of metering instruments. It is necessary for a reliable operation of relays that, at short-term short circuits, changes in a wide range of current intensity are proportional to the changes in the primary current intensity. The situation is opposite in the case of metering instruments for which it is desirable that a short-circuit secondary current increase is as low as possible. Therefore, current transformers with two secondary circuits are often used. Protective instruments are connected to one of them and measuring instruments to the other one. Usually, current transformers have the rated secondary current intensity of 5 A.

Both voltage and current transformers have a similar design. They have a steel core with high magnetic conductivity and copper primary and secondary winding coils wound on the core. The whole is often not enclosed and is secured directly to the conductors of the EAF power supply system.

The EAF, as in any other electric system, is equipped with automatic instruments protecting against failures and deviations from normal operating conditions. The protecting instruments should trip the furnace power system if operating conditions indicating the possibility of damage occur. In many cases, when the ensuing new operating conditions do not result from power system damage, but are rather caused by an incorrect operation (e.g. furnace operation at a current intensity higher than the allowed limit), it is not necessary to shut off the furnace immediately. In a case like this, correct current operating conditions will be established automatically or the protections will generate an appropriate signal to inform the operators about the occurring malfunctioning. It is only when the automatic controller or operators do not remove the existing anomaly in the furnace operation that the furnace will be disconnected from the grid by the protection system.

The furnace is automatically disconnected from the grid by a special protection relay system. Each type of failure is protected by a separate relay. Current and voltage relays are most often used in the EAF protection system. In terms of design, there are the following types: magnetoelectric, electrodynamic, and inductive. Relays have their own current characteristic, where the sizes of the necessary pulse (voltage or current intensity) and on-time are distinguished. In addition, they should have adequate accuracy, selectivity, and reliability in operation. Relay selectivity means the adaptation of the relay to shut off only the power or control system circuits that were damaged or whose operation parameters exceeded the acceptable values and cannot be self-adjusted. The necessary selectivity is obtained by setting correct pulse sizes and relay start-up times. As a rule of thumb, the closer a relay to the source of electrical energy use, the shorter its start-up time.

The surge protection, or the protection of the furnace power system against a rapid increase in the supply voltage, is very important. The research of arc furnace power systems showed that when switching off the transformer, which results in a sudden loss of current in the secondary circuit, voltages may be induced in the circuit, even 10 times higher than the rated voltage. Therefore, protections against negative effects of such high voltages are applied in EAF power systems. To this end, capacitor batteries with a specific capacitance are installed between the oil circuit breaker and the transformer primary winding, parallel to the latter. The value of voltage induced during a transformer shutoff is then determined from the equation:

$$U = I\sqrt{\frac{L}{C}} \tag{4.7}$$

where:

U – voltage, V,
I – current intensity, A,
L – inductance of the transformer secondary winding, H,
C – capacitance of the capacitor battery, F.

It appears from the above equation that the voltage induced during a transformer shutoff can be reduced by increasing the system capacitance, that is, by connecting a battery of capacitors. In practice, depending on the electrical parameters of the supply system, the capacitance of such a battery is specified, and it usually does not exceed a few microfarads.

4.6 INDUCTION STIRRER

Induction stirrers are in fact sometimes applied in EAFs, but rather in older design facilities and in small furnaces. In this case, the stirrer's purpose is to intensify the metal bath stirring process, in particular when high-alloyed steels are manufactured, where substantial amounts of sparingly soluble alloy additions need to be added to the melt. In larger furnaces, at high transformer powers, the problem of an adequate melting rate and subsequent melt stirring is less relevant. Therefore, in such furnaces, stirrers are less frequently used.

The principle of induction stirrer operation is similar to the operation of a stator of a two-phase asynchronous motor, for which the metal bath becomes a shorted rotor. The stirrer coil winding is made of copper conductors insulated with mica and a fiberglass tape. The whole is cooled with water. The stator's core and windings as well as the water-cooling pipes are located under the nonmagnetic bottom of the furnace shell inside a special protective cover. Only four terminals for connecting power lines, connector pipes for the cooling water supply and discharge, and terminals of the temperature measurement system go outside. In order to protect the stirrer against damage, the cooling water temperature is measured at a few points with resistance thermometers. The method of the stirrer location and connections of its coils are shown in Figure 4.9.

Induction stirrer coils are supplied with a low-frequency current between 0.3 and 1.5 Hz. The transducer power and the supply current intensity are adjusted to the furnace capacity and hearth thickness. In order to obtain a rotating electromagnetic field, the stirrer stator winding consists of two coils, which are supplied with currents

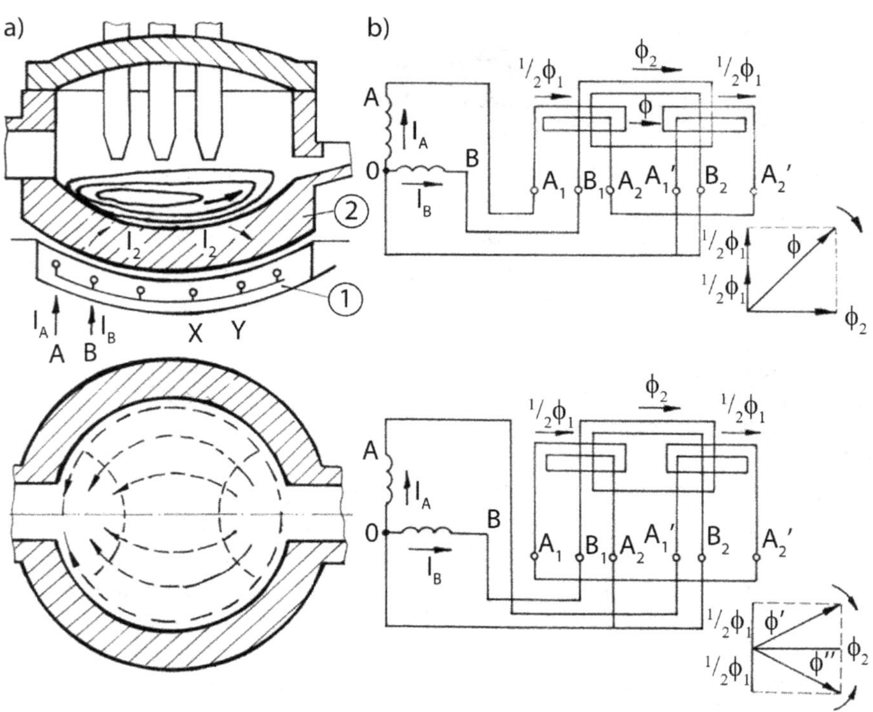

FIGURE 4.9 Diagram of induction stirring of a metal bath in an electric arc furnace with a stirrer. (a) Method of stirrer fixing. (b) Stirrer design: 1 – induction stirrer, 2 – furnace hearth. (Author's work.)

shifted at a phase angle of 90°. The direction of the electromagnetic field rotation, e.g. from the main door to the taphole or vice versa, "converging" to the furnace axis, or "diverging" from the furnace axis, can be reversed by changing over the connection of the coil windings. The electromagnetic field induces eddy currents in the melt, and by interacting with the rotating electromagnetic field of the stator, the eddy currents cause the movement of the liquid metal layers adjacent to the hearth. Metal, flowing along the hearth, reaches the surface of the melt and subsequently goes in the opposite direction, causing the movement of its outer layers. The metal movement in the horizontal plane is obtained in a similar manner.

To expedite the dissolution of alloy additions in the metal bath and to make its chemical composition and temperature more uniform, the metal movement directions in the vertical and horizontal planes are cyclically changed. The metal movement speed depends on the current frequency and the stirrer power. The depth of penetration of the electromagnetic field force lines into the metal bath is inversely proportional to the square root of current frequency and it can be determined from the equation:

$$\delta = \frac{59-62}{\sqrt{f}} \tag{4.8}$$

where:

δ – penetration depth, cm,

f – current frequency, Hz.

The operational data show that the application of an induction stirrer allows a better uniformity of chemical composition and temperature of the metal bath to be obtained. In addition, the tap-to-tap time is shortened and the level of the energy and electrode consumption is reduced. It is particularly noticeable when high-melting sparingly dissolving alloy additions are used for the heat, such as ferrochromium, ferromolybdenum, and ferrotungsten. If there is no stirrer, these alloys settle at the bottom, where the temperature is lower, and they dissolve very slowly.

A negative effect of the use of an induction stirrer may be a greater wear of the refractory lining of the furnace hearth and a possible increase in the content of exogenous nonmetallic oxide inclusions in the steel produced. Increased erosion of the hearth refractory lining is related to a higher temperature of the metal touching the hearth and a higher speed of its flow.

In large furnaces, with a high transformer installed capacity, and using lances for gaseous oxygen blowing, stirring of the metal bath is sufficient without stirrers. Therefore, the furnaces of this type are usually not equipped with induction stirrers.

4.7 OPTIMAL OPERATING CONDITIONS OF THE EAF POWER SUPPLY SYSTEM

During the steelmaking process in EAFs, the electrode position height and power input need to be controlled. The furnace is started up with automatic controllers on and raised electrodes. Automatic controllers lower the electrodes until they touch the charge. When two electrodes touch the charge, the so-called process short circuit occurs, the contact place is heated by resistance, the electrodes raise automatically, and arcs start burning. The process of scrap meltdown starts, with heat from the arc radiation energy. Melting scrap in liquid form flows down, causing the need to lower the electrodes to maintain the arcing. Due to a small volumetric density of scrap and its nonuniform distribution, during the meltdown, short breaks in the arc burning or process short circuits often appear, and therefore, the electrode position height must be continuously adjusted automatically in order to maintain a proper length of the electric arc.

The amount of electric energy converting into heat energy in the arc depends on the electric parameters of the power system and the parameters of the arc itself. For cost-efficiency, the share of electrical energy converted into heat energy should be as high as possible, as only this part of energy is used for the execution of the metallurgical process and covering heat losses accompanying this process. The electricity consumption can be optimized and the furnace stable operation can be achieved by selecting the correct electric parameters of the arc, and applying fast responding automatic controllers for its control.

4.7.1 Characteristics of the Electric Arc

The electric arc is a form of electric discharge, where the electric current is conducted by ionized gases, metal or oxide vapors occurring in the space between the electrodes. To initiate the arc, the electrodes should be shorted to heat the place of their contact. At that time, a high temperature appears at the contact place, and causes thermal ionization (the so-called thermionic emission), which enables free electrons to appear at this place. Then, the parting of the electrodes, which are connected to the power supply, should start slowly. Subsequently, an electric field is generated between the live electrodes, and the electrons in this area are attracted to the positive electrode. Electrons, moving at very high speeds and colliding with atoms from the space between the electrodes, cause the collision ionization of these atoms. As a result, an ionized stream of matter forms between the electrodes, called plasma (a channel of conductive electrons and ions), enabling the electric current to flow between the electrodes. The whole effect is called an electric arc.

As the distance between the electrodes increases, the resistance, arc power, and its temperature increase. The arc between the electrodes will exist when the electrodes are supplied either with direct or alternating current. The diagram of an electric arc discharge is shown in Figure 4.10.

For an EAF, the arc burns between a graphite electrode and pieces of scrap or liquid metal bath. A moving so-called arc root, from which the flux of electrons is emitted, is located at the face of a graphite electrode. A so-called arc column is distinguished under the root. Inside the arc column, concurrently with ionization, the effect of recombination and diffusion, which is reconnecting free electrons with cations, occurs. In favorable conditions, electrons can combine with neutral particles, forming negatively charged anions, which subsequently potentially can recombine with cations. The arc column is surrounded by a flame, consisting of hot gases moving toward colder zones [10].

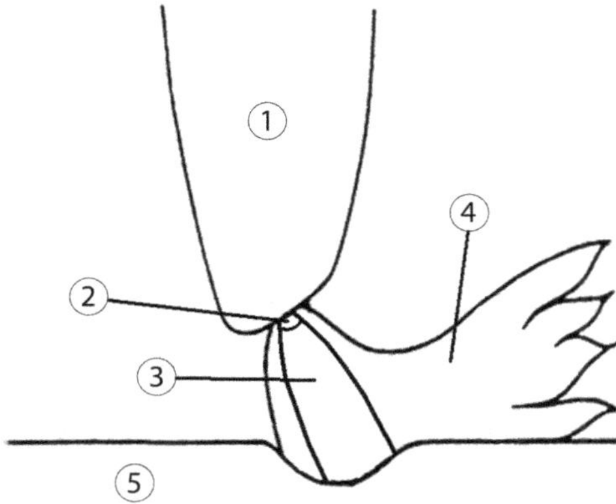

FIGURE 4.10 Diagram of an electric arc discharge: 1 – electrode, 2 – arc root, 3 – arc column, 4 – flame, 5 – material (scrap, metal bath). (Author's work.)

The temperature in the arc root can achieve values close to the temperature of graphite sublimation, it is estimated at 3,873–4,573 K (3,600– 4,300°C), and the amount of energy emitted in this area of the arc accounts for approximately 7%–10% of the total arc energy. A substantial part of this energy is transferred to the electrodes as a result of conduction.

The arc column temperature is much higher, and it also features very steep gradients on the cross section. For the arc diameter of 20–50 mm, it is estimated that the temperature in the arc axis can reach the values between 20,000 and 30,000 K. However, the temperature at the outer surface of the arc column is from 6,000 to 7,000 K. The average temperature of the arc column can be estimated from the following equation:

$$T = 800\ U_i \tag{4.9}$$

where:

U_i – ionization potential of the gas in which the arc is burning, V.

Over 90% of the total energy of the arc is emitted in the column. Over half of this energy is transmitted as a result of conduction and radiation directly to the scrap or melt. The balance is transferred by radiation and convection to the space around the column, most of it by flames. This part of energy pervades to the scrap or melt, electrodes, and slag as well as heats the inside of the furnace, including its walls and roof.

During the EAF operation, in particular at the beginning of the scrap meltdown, the conditions for the burning arc are unstable. The conducting channel length constantly changes due to the scrap movements caused by its melting. More stable conditions for the burning arc emerge after the meltdown, when the arc is burning toward the surface of the liquid metal bath. In this case, the thermal conditions stabilize, and the processes of ionization and recombination in the plasma channel proceed at the same rates. At that time, the current intensity curve has no breaks and smoothly goes through the zero value when the supply current changes its sign (sinusoidal). If the thermal conditions of the arc operation are disturbed, the overdeveloped ion recombination can cause the effect of arc breaking (its unstable operation). The current density in the arc column is from a few hundred to a few thousand amperes per 1 cm^2 of the cross-sectional area.

For a sinusoidal alternating current, existing in the classic power supply grid, during each semi-period the voltage and current intensity reach values equal to zero (Figure 4.11a–c). In a burning electric arc (without inductance in the circuit), when the supply voltage approaches zero, the arc goes out for a moment, and only reappears when the voltage achieves the value U_1 (Figure 4.11a). At this moment, a current intensity of I_1 appears in the circuit, and its value changes in accordance with a sinusoidal curve, but with periodic breaks. The arc voltage value is constant and it is U_2. The arc goes out when the supply voltage achieves U_2, that is, before the zero voltage is achieved. Next, the whole effect is repeated periodically, but in this case, the arcing is unstable (with breaks). The time of breaks in the arc operation depends on the properties of graphite from which the electrode is made and the temperature. A stable operation of an alternating current arc is obtained between two graphite electrodes (low exit work of electrons), or between a graphite electrode and the metal charge heated to a high temperature. The inductance of the electric circuit in which the arc burns is important. The inductive resistance in the circuit causes a vector shift of the current relative to the voltage, which causes an "extension of the existence time" of the arc voltage U_2 and a change in the nature of current intensity curve, as presented in Figure 4.11b. Then, both the voltage and the current intensity are continuous, without breaks. The practical curves of arc voltage and current intensity, measured in industrial conditions, are shown in Figure 4.11c.

During the steelmaking process in an EAF, a cold charge is used and, therefore, to achieve a stable operation of the electric arc from the beginning of the melting, it is necessary to apply inductive resistance in the power supply circuit. If the reactance

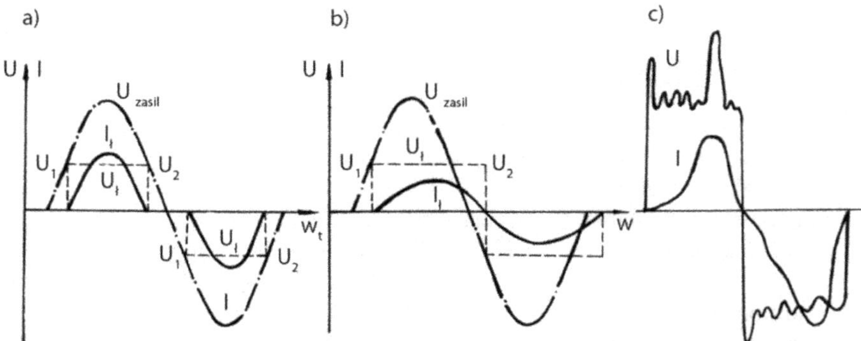

FIGURE 4.11 Changes of electric arc voltage and current intensity. (a) Without inductance in the circuit. (b) With inductance in the circuit, (c) measured in industrial conditions. (Author's work.)

of the power system is too low (small furnaces), a reactor with an induction winding is added to its circuit. For large furnaces, the power system reactance is sufficient and electric arcs operate in a stable manner without additional inductance.

The relationship between the current intensity, voltage, and resistances in the electric arc circuit can be analyzed on the basis of an equivalent circuit shown in Figure 4.12. Relationships between the parameters of a circuit like this are described by Ohm's law:

$$I = \frac{U}{\sqrt{R_c^2 + X_t^2}} = \frac{U}{(R_t + R)^2 + (\omega L)^2} \quad (4.10)$$

where:

U – supply voltage (secondary voltage of the furnace transformer), V,
I – electric arc current intensity, A,
R_t – power supply system resistance, Ω,
R – electric arc resistance, Ω,
R_c – overall resistance, Ω,
X_t – power supply system reactance, Ω,
ω – frequency, Hz,
L – power supply system inductance, H.

The electric arc voltage is proportional to the arc length, which is expressed by the dependence:

$$U_a = \alpha + \beta l \quad (4.11)$$

where:

U_a – arc voltage, V,
α – sum of voltage drops in the electrode and the metal charge, V,
β – unit voltage drop in the arc column, V/mm,
l – length of electric arc, mm.

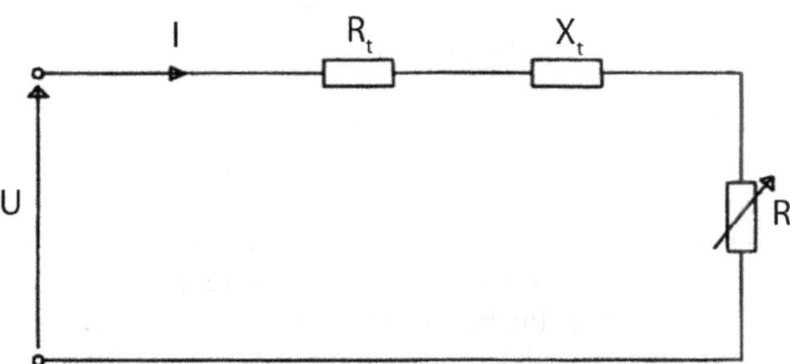

FIGURE 4.12 Equivalent circuit of the electric arc furnace power supply system. (Author's work.)

A proportional relationship between the arc voltage and length only occurs within the range of their values corresponding to the stable arc operation. If the electric arc is considered as a resistor connected in series in the furnace power supply system, its voltage can be determined from the equation:

$$U_a = \frac{U \cdot R}{(R + R_t)^2 - X_t^2} \tag{4.12}$$

Equations (4.10)–(4.12) provide a dependence describing the relationship between the electric arc length and voltage as well as the resistance and reactance of the furnace system. This dependence has the following form:

$$l = \frac{1}{\beta}\left(\frac{U \cdot R}{\sqrt{(R + R_t)^2 - X_t^2}} - \alpha\right) \tag{4.13}$$

This equation implies that the arc length increases as the supply voltage increases, the furnace system resistance and reactance decreases, and the β value decreases and the α value increases. The α value depends on the electrode, metal charge, and slag properties. For the graphite electrode-steel scrap system $\alpha = 22$, and for the graphite electrode-basic slag system $\alpha = 9$. The β value depends on the melting phase. During meltdown, it is 10–12 V/mm, and the electric arc is short. In the refining period $\beta = 1.1$ V/m, and the arc extends.

The electric arc generates its own magnetic field around, which causes its spiral swirl and circular movement. At the same time, the impact of magnetic fields from other arcs causes a deviation of its burning direction from the plumb-line, outward, toward the walls [10]. The arc also swirls on the electrode face, which is related to the wear of the electrode tip. The arc diameter is much shorter than the diameter of the electrode face. During burning, the arc "stops" at the lowest spot of the electrode tip, at the shortest distance to the charge or the metal bath (the arc always "chooses" the shortest distance between the materials to where it burn). All of the described effects change very dynamically, which causes very fast changes of the arc position in relation to the electrode face. The classic electrode tip shape, including the most frequent position of the arc, is shown in Figure 4.13.

During arcing in real conditions, changes in the voltage and current intensity of the arc are not sinusoidal, which is related to the nature of the arc itself and external disturbances. These disturbances include scrap melting, a change in the arc length during its jump from a scrap piece onto another piece, the wavy motion of the metal bath surface, etc. In addition, the saturation of the transformer core and other devices, which disturbs the proportionality between the magnetizing current and the magnetic field, influences the nonsinusoidal shape of these curves. All of these effects cause a number of detrimental impacts on the power supply system operation, including an increase in the electric loss in the power supply system.

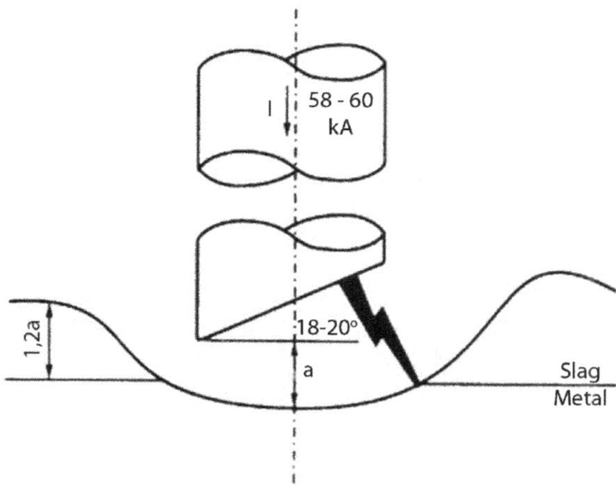

FIGURE 4.13 View of an electrode tip with a burning arc. (Author's work.)

4.7.2 Power Controllers for the EAF Power Supply System

The power input is controlled in EAFs by changing the secondary voltage of the furnace transformer or by changing the resistance (length) of the electric arc. The primary winding taps of the furnace transformer as well as the arc length are controlled by power controllers, performing the task of maintaining a constant resistance of the electric arc. The control signal is obtained by measurements of the electric arc current intensity and voltage. The purpose of the control system is to maintain such a distance between the electrode and the solid charge or liquid melt, at which the optimum energy consumption and the optimum impact of the arc on the metal bath are obtained. A structural diagram of the power automatic control system is shown in Figure 4.14.

Each phase of the furnace power supply system is an individual controlled object. Each phase has its own independent power control system (controller). The controller comprises an array of devices allowing the measured value to be compared with the set value and the occurring deviations from the set values as well as influencing the arc length to be eliminated. The following are the primary components of the control system: the controlled object, measuring and comparing element, amplifier, actuator, and reference input element. The foregoing elements are included in all of the control systems. The following elements complete and stabilize the operation of controllers: negative feedback between the actuator and comparator, negative feedback between the amplifier and comparator, feedback element, and the system of additional connections between the controlled system and the comparator, used for changing the results of comparisons at a change in the furnace transformer secondary voltage. These elements are only used in expanded controllers.

The measuring and comparing elements contain instruments for measuring and comparing the controlled variables with the set values. The signal, proportional to

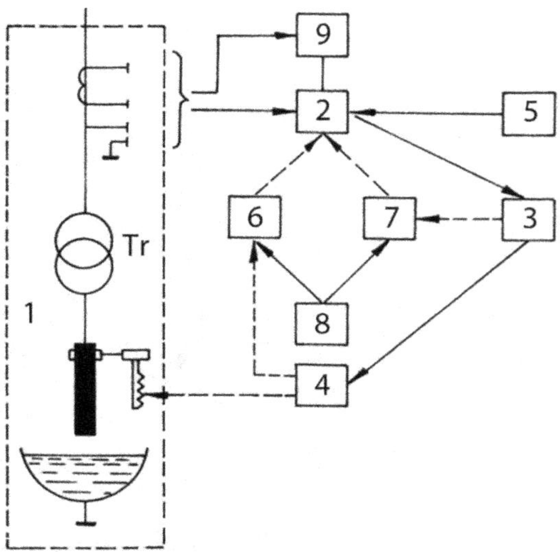

FIGURE 4.14 Structural diagram of the power automatic control system of an electric arc furnace: 1 – EAF with its power supply system (controlled object), 2 – measuring system and comparator, 3 – amplifier, 4 – actuator, 5 – reference input element, 6 – negative feedback between the comparator and the actuator, 7 – negative feedback between the amplifier and comparator, 8 – feedback element, 9 – system of additional connections between the controlled system and the comparing system. (Author's work.)

the difference between the compared values, is fed to the amplifier, and after amplification, it goes to the actuator. The actuator causes a shift of the electrode, allowing the error to be eliminated. Elements 2–4, along with the controlled system, constitute a closed loop, where the control signals flow in one specific direction.

Element 7 changes the comparison results between elements 3 and 2 depending on the obtained values of current intensity on the amplifier output. Element 6 changes the comparison results between elements 4 and 2 as a function of the electrode shift speed. Element 9 makes an additional connection between 1 and 2, changing the results of comparison depending on the voltage step. This design of the structural diagram of power control in an EAF with complementary and stabilizing elements is necessary due to its adaptation to various complex needs that exist in steelmaking plants.

On the basis of many years of experience in operation of EAFs, the following requirements are set to power controllers:

- sensitivity regarding the controlled variable should protect the furnace operation at an acceptable range of deviations (e.g. the dead zone during melting period can be ±(3%–6%), and in other periods of melting ±(2%–4%),
- the system speed should allow the maximum disturbances with the aperiodic nature of control to be eliminated not later than 1.5–3.0 s. Start-up and braking time of the controller should be between 0.2 and 0.3 s,

- limit speed of the electrode shift should be 4.5–6.0 m/min,
- variable change of the set furnace power should be possible within the range 20%–125% of the rated power, with the accuracy not lower than 5%.
- automatic arc ignition should be possible and stopping all furnace electrodes when the supply voltage fades,
- simple design and should be reliable in operation.

In older solutions, electromachinery power controllers were applied, sometimes hydraulic controllers. Now, as a result of high-power electronic technologies, thyristor or inverter power controllers are applied for electrode position control, including control of arc electric parameters, which also enable all these variables to be globally optimized [11–14].

4.7.2.1 Thyristor Power Controllers

Thyristor systems meet the requirements concerning controllers. A schematic diagram of a system like this is shown in Figure 4.15. The controller consists of the following blocks: power controller, dead zone control, feedback limitation, power supply, amplifiers, power supply and control, pulse-phase thyristor control, thyristor transformation, and drive motor excitation. A signal proportional to the arc voltage and current is formed in the power controller. This signal, which determines the rotational speed of the motor *S* of the electrode shifting mechanism, is transmitted by the dead zone control block to the input of semiconductor amplifiers. A signal in the form of

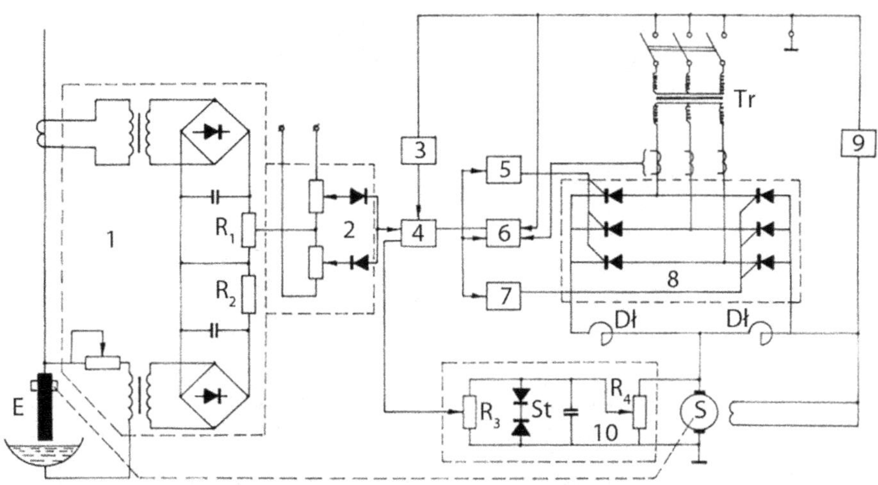

FIGURE 4.15 Schematic diagram of a thyristor power control system of an electric arc furnace: 1 – power controller, 2 – element for dead zone control, 3 – system power supply block, 4 – amplification block, 5 – thyristor pulse-phase control block, 6 – thyristor system control block, 7 – thyristor pulse-phase control block, 8 – thyristor assembly, 9 – motor excitation power supply block, 10 – block of feedback between the element for dead zone control and the actuator, S – motor driving the electrode feed, as the actuator. (Author's work.)

negative feedback is also supplied from the voltage on the rotor of motor S. From the amplifier output 4, the signal is fed to the block of pulse-phase device 5 or 7. The mean value of the "rectified" voltage and the rotational speed of the motor are controlled by the phases of change of the thyristor firing angle in block 8. The feedback signal at the electrode movement upward is limited to a specific pre-set level in block 10.

If there are no deviations from the set operating conditions, there is no signal from the power controller block 1. The output voltage of summing amplifier 4 is equal to zero, and the actuator, which is motor S, is not working. After exceeding the set furnace operating conditions, a signal appears at the output of block 2, and the sign of this signal depends on the nature of the disturbance (in a boundary case, it is an arc break or a short circuit). When the signal value exceeds the dead zone area, voltage appears at the amplifier output downstream the thyristor pulse-phase control block. One of the blocks 5 or 7 starts "closing", and then voltage appears at its output. Due to this voltage, one of the complexes of the thyristor transformation block starts "opening" as a result of decreasing its control angle. At the same time, the other one of the blocks 5 or 7 is "open" and the control angles in the other complex of thyristor transformation are at their maximum. The motor of electrode moving gear starts accelerating, and the feedback voltage input from resistor R_3 increases.

For a small signal from the power controller, the rotational speed of the motor *S* is determined by the difference between this signal and the feedback from the rotor voltage: in this case, the system works within the proportional control range. For a large signal from the power controller, the feedback signal is limited by the electronic voltage stabilizer S_t, and the controller works in the nonproportional range, which is necessary for the fast elimination of large disturbances. The controller dead zone is set with resistors R_1 and R_2, and the range of proportional control is set with resistors R_3 and R_4. Thyristor controllers are mostly used for work in smaller sized furnaces.

4.7.2.2 Hydraulic Power Controllers

The application of a hydraulic drive allows the electrode movement design to be significantly simplified. The gears are much lighter, smaller, and easy to control. The hydraulic drive is fast responding, flexible, and accurate. The controllers are used not only for the power control of a hydraulic actuator but also for an amplifying system. Only the measuring and comparing systems are electrical. A structural diagram of a hydraulic power controller is shown in Figure 4.16.

On one hand, a signal proportional to the current flowing through the current transformer 1, controller 2, and the dividing transformer 3 is transmitted to the excitation and control coils of motor 4. On the other hand, the signal coming from the arc voltage and input from the dividing transformer 5 is transmitted to the excitation and control coils 6. Capacitor 7 protects the phase shift between the excitation and control currents, which is necessary for the torque to occur:

$$M = K \cdot U_1 \cdot U_2 \cdot \sin\varphi \tag{4.14}$$

where:

K – unit conversion factor,

U_1 – voltage on the excitation coil,

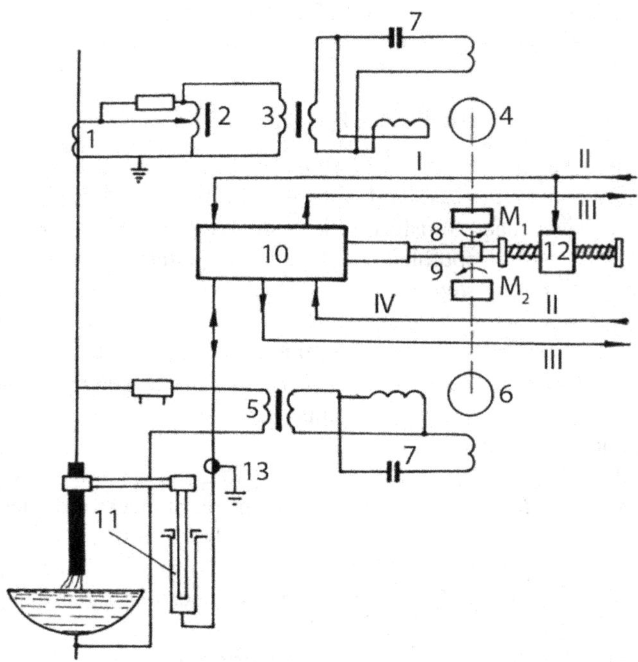

FIGURE 4.16 Schematic diagram of a hydraulic power control system of an electric arc furnace: I – hydraulic control system, II – hydraulic electrode lifting system, III – hydraulic electrode lowering system, IV – central hydraulic system, 1- current transformer, 2 – master controller, 3 – transformer, 4 – motor, 5 – transformer, 6 – motor, 7 – capacitor, 8,9 – mechanical drive assembly, 10 – hydraulic amplifier, 11 – hydraulic cylinder piston, 12- reactor, 13 – hydraulic cylinder piston shift controller. (Author's work.)

U_2 – voltage on the control coil,

φ – phase shift angle between the control coil current and the excitation coil current.

If there are no deviations from the set operating conditions, the torques of motors 4 and 6 are equal, and because they have opposite senses, the system is at a standstill. If this balance is disturbed, the torque of one of the motors will be higher and then the controller will be enabled. The pinion 8 installed on the motor shaft engaged with the rack 9 causes a shift of a hydraulic slide of the amplifier 10. This causes the movement of the cylinder 11 piston in the right direction. The reactor 12 protects the control system against reacting to momentary deviations from the set furnace operating parameters. The hydraulic power controllers are controlled with booster pumps with a fluid pressure up to 1,200 kPa. The electrode lifting gear cylinder is supplied from the central hydraulic mains. The maximum speed of electrode movement is controlled with the controller 13.

Hydraulic power controllers feature a high response speed, and high sensitivity at the electrode lifting speed up to 5 m/min. The dead zone as regards the current intensity is ±3%; the time constant is 0.035 s, and the inertia time is 0.08 s. They have proportional characteristics within the range of 80%–100% of the current intensity.

4.7.3 Optimization of the Power Input

The efficiency of EAFs and electricity consumption per metric ton of steel manufactured largely depends on the ability to use the installed capacity of the power supply system. The furnace transformer is designed to enable various values (steps) of secondary voltage to be obtained by adjusting the number of turns in the primary winding. For each voltage step, the optimum value of electric arc power, electrical efficiency, and power factor cos φ can be obtained only at strictly specified values of electric parameters. The value of these parameters for each voltage step is defined on the basis of the furnace electrical characteristics that can be determined with analytical methods or on the basis of measurements of the transformer operating parameters with no load and at short-circuit currents.

An equivalent diagram of the EAF power system for one phase is presented in Figure 4.17. Symbols R_{t1} and R_{t2} mean equivalent resistances of devices connected in series of the primary and secondary side, respectively. Similarly, symbols X_{t1} and X_{t2} mean equivalent reactances of devices connected in series of the primary and secondary side, respectively. Symbol R means the electric arc resistance, subject to the variability of this value.

Impedances on the primary side can be converted into the secondary side. Then, we obtain an equivalent diagram of one phase of the power supply system as shown in Figure 4.12. Symbol U means the phase voltage of the transformer secondary side, R_t and X_t equivalent resistance and reactance of the whole furnace power supply system. The intensity of the current flowing in the transformer secondary circuit, furnace secondary circuit, and the electrodes is identified with symbol I.

Each EAF has a strictly determined value of resistance and reactance of its power system, which can be established by the appropriate measurements. In accordance with the diagram shown in Figure 4.12, the value of current intensity of the arc is defined by the equation:

$$I = \frac{U}{\sqrt{(R_t + R)^2 + X_t^2}} \tag{4.15}$$

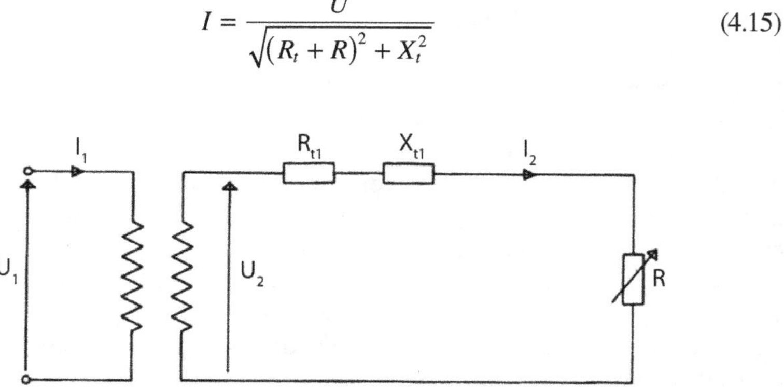

FIGURE 4.17 An equivalent diagram of an electric arc furnace power supply system. (Author's work.)

It can be seen from equation (4.15) that electric arc resistance is the basic variable in the power supply circuit of the furnace. Depending on the arc resistance, the current intensity changes as well as other electrical parameters dependent on changes in the current intensity. The relationship between the current intensity and individual types of power is defined by the following equations [15]:

$$S = 3 \cdot 10^{-3} \cdot U \cdot I \tag{4.16}$$

$$P = 3 \cdot 10^{-3} \cdot I^2 \sqrt{\frac{U^2}{I^2} - X_t^2} \tag{4.17}$$

$$Q = 3 \cdot 10^{-3} \cdot I^2 X_t \tag{4.18}$$

$$P_e = 3 \cdot 10^{-3} \cdot I^2 R_t \tag{4.19}$$

$$P_a = P - P_e = 3 \times 10^{-3} \cdot I^2 \left(\sqrt{\frac{U^2}{I^2} - X_t^2} - R \right) \tag{4.20}$$

$$\eta_{el} = \frac{P_a}{p} \tag{4.21}$$

$$\cos\varphi = \frac{P}{S} \tag{4.22}$$

$$U_a = \frac{10^3 \cdot P_a}{3 \cdot I} = \sqrt{U^2 - (I \cdot X_t)^2} - I \cdot R_t \tag{4.23}$$

where:

- S – furnace apparent power, MVA,
- U – supply transformer secondary side voltage, V,
- I – current intensity in the transformer secondary circuit, kA,
- P – furnace active power, MW,
- X_t – equivalent reactance of the furnace power supply system, Ω,
- Q – furnace reactive power, MVAr,
- P_e – electric losses power, MW,
- P_a – electric arc power (active), MW,
- $\cos\varphi$ – power factor,
- η_e – furnace electric efficiency,
- U_a – arc voltage, V.

It appears from the aforesaid dependences, assuming constant values of R_t and X_t, that the current intensity values can be calculated as a function of the arc resistance

R for each voltage step, and subsequently the other characteristics of EAF operation can be derived. The above characteristics can be plotted as a function of current intensity in the secondary circuit [15]. The graphs of apparent power, reactive power and active power in the secondary circuit of the furnace supply system, arc power, and electric losses power are shown in Figure 4.18. Figure 4.19 shows a graph of the electric efficiency factor and power factor cos φ.

As can be seen in equations (4.17) and (4.20) as well as the curves in Figure 4.18, the active power in the secondary circuit of the furnace supply system and arc power are functions of current intensity, having a maximum. As the current intensity increases, initially the powers increase until they reach the maximum and then they decline. For the short-circuit current I_z, the arc power is zero, and the active power is equal to electric losses power. The value of the short-circuit current can be determined on the basis of Ohm's law for the equivalent system of EAFs (Figure 4.12):

$$I_z = 10^{-3} \frac{U}{R_t^2 + X_t^2} \tag{4.24}$$

The current intensities at which the maximum active power and the maximum arc power occur are determined from equating the derivatives of functions (4.17) and (4.20) to zero. Then, we obtain:

$$I_2 = 10^{-3} \frac{U}{\sqrt{2} \cdot X_t} \tag{4.25}$$

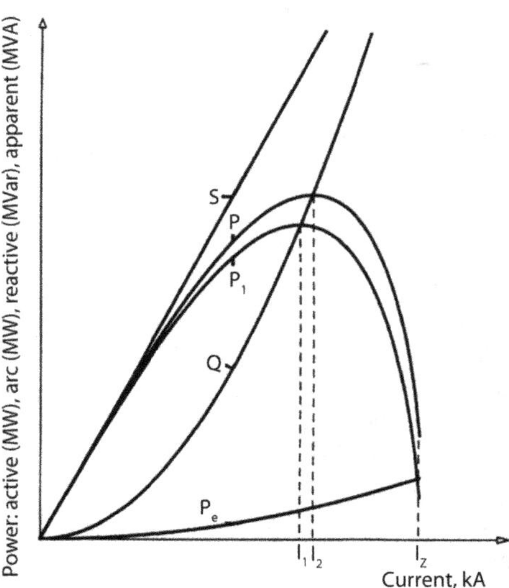

FIGURE 4.18 Powers curves for electric arc furnace operation as a function of current intensity in the secondary circuit (apparent power, active power in the secondary circuit of the furnace supply system, arc power, reactive power, and electric losses power). (Author's work.)

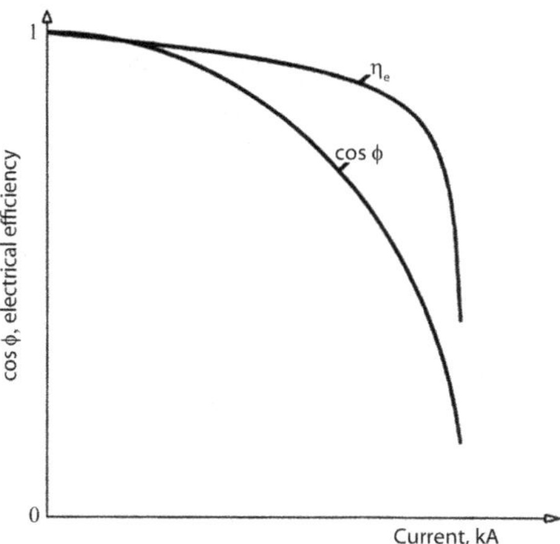

FIGURE 4.19 Power factor and electric efficiency factor curves for electric arc furnace operation as a function of current intensity in the secondary circuit. (Author's work.)

$$I_1 = 10^{-3} \frac{U}{2\sqrt{R_t^2 + X_t^2}\left[\sqrt{R_t^2 + X_t^2} + R_t\right]} \tag{4.26}$$

where:

I_1, I_2, I_z – current intensities of the furnace power system secondary side for the maximum arc power, maximum active power, and short-circuit current, respectively, kA,

U – supply transformer secondary voltage, V,

R_t, X_t – furnace power supply system equivalent resistance and reactance, Ω.

It can be seen in Figure 4.18 that the value of current intensity I_1 at which the maximum electric arc power occurs is lower than the value of current intensity I_2 at which the maximum active power occurs. Note that this relationship applies to each EAF power system. The obtained values of the maximum active power and arc power, respectively, are:

$$P_{a\max} = \frac{3 \times 10^{-6} \cdot U^2}{2\left[\sqrt{R_t^2 + X_t^2} + R_t\right]} \tag{4.27}$$

$$P_m = \frac{3 \cdot 10^{-6} \cdot U^2}{2 \cdot X_t} \tag{4.28}$$

where:

$P_{a\max}, P_m$ – maximum arc power and maximum active power, MW.

From the analytical calculations and Figure 4.18, it appears that, for each furnace, at the current intensity of I_2, the active power equals the reactive power, whereas $\cos\varphi = \sqrt{2}/2$.

The operating characteristics of a steelmaking EAF are plotted in a similar way. To build these curves, it is necessary to know the furnace heat losses and unit consumption of electrical energy. The actual values of heat losses and unit electricity consumption can only be determined on the basis of data from the operation of an EAF. At the designing stage, the above-mentioned values can only be estimated.

Heat losses depend on the type and thickness of the refractory lining (for classic refractory lining) or the water-cooling system design, if applicable. For a classic basic refractory lining, the furnace heat losses can be estimated on the basis of empirical formulas [15]:

- for furnaces with a capacity under 10 tons

$$P_{sc} = 0.24\,G^{0.4} \tag{4.29}$$

- for furnaces with a capacity from 10 tons to 30 tons

$$P_{sc} = 0.25\,G^{0.4} \tag{4.30}$$

- for furnaces with a capacity over 30 tons

$$P_{sc} = 0.26\,G^{0.4} \tag{4.31}$$

where:

P_{sc} – furnace heat losses, MW
G – furnace capacity, ton

Relationships (4.29)–(4.31) concern refractory lining immediately after repair and a few first heats in a campaign. As the furnace refractory lining campaign passes, the heat losses increase. In the final stage of the campaign, the heat losses are up to 50% higher than the value established with the above relationships. For acid refractory lining, the heat losses are about 20% lower.

The unit electricity consumption depends on the furnace design, grades of steels manufactured, the technology applied, and it differs in the subsequent stages of melting. The unit electricity consumption for operating furnaces ranges from 400 to 900 kWh/t of steel manufactured. These values concern the case when no additional heat sources are used during melting in an EAF. Such sources can include the use of preheated scrap as a charge, use of gas-oxygen or oil-oxygen burners during melting, and injection of carbon-bearing materials and oxygen. It can lead to a decrease of the unit electricity consumption, even by half. The theoretical value of energy needed to heat and melt 1 ton of steel depends on its chemical composition and is 340–370 kWh/t [15].

On the basis of the assumed aforesaid values, you can write formulas to determine the operating characteristics:

$$w = e_t\left(1 + \frac{P_e + P_{sc}}{P_a - P_{sc}}\right) \tag{4.32}$$

$$\eta = \frac{P_a - P_{sc}}{p} \tag{4.33}$$

$$t = \frac{e_t \cdot G}{P_a - P_{sc}} \cdot 10^{-3} \tag{4.34}$$

$$g = \frac{P_a - P_{sc}}{e_t} \cdot 10^{3} \tag{4.35}$$

where:

w – unit electricity consumption for melting 1 ton of steel, kWh/t,
e_t – theoretical amount of energy necessary to melt down 1 ton of steel (340–370), kWh/t,
P_e – electric losses power, MW,
P_a – arc power, MW,
P_{sc} – heat losses, MW,
P – active power, MW,
η – overall efficiency of the EAF,
t – charge meltdown time, h,
g – charge meltdown unit productivity, h.

The electric losses power, arc power, and active power of the EAF are functions of secondary current intensity. Therefore, all of the operating characteristics (4.32)–(4.35) can be expressed as functions of current intensity as shown in Figure 4.20. As can be seen from the curves, the minimum time of charge meltdown and the maximum unit efficiency of charge meltdown are obtained for the current intensity I_1 or at the maximum arc power. The minimum unit electricity consumption for melting 1 ton of steel is obtained at the maximum overall furnace efficiency. The maximum overall furnace efficiency is obtained for the current intensity of:

$$I' = \sqrt{\frac{U^2 \cdot P_{sc}}{3U^2 \cdot R_t + 2X_t^2 \cdot P_{sc} \cdot 10^6}} \tag{4.36}$$

where:

I' – current intensity, at which the maximum overall furnace efficiency occurs, kA,
(other designations as above).

It appears in Figure 4.20 that the value of current intensity I' is lower than the value of current intensity I_1. This is the case for each EAF power supply system. As

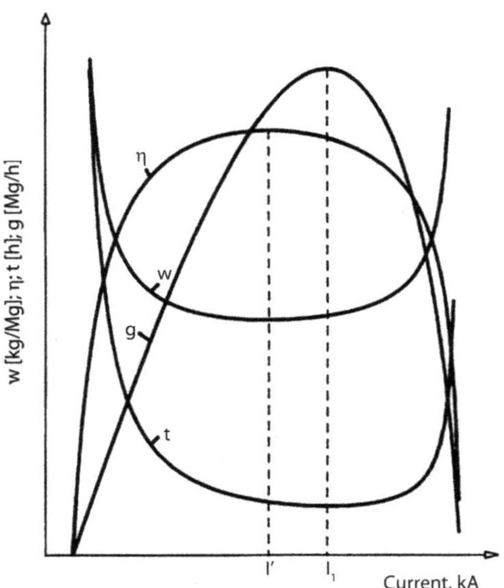

FIGURE 4.20 Characteristic curves for electric arc furnace operation (curves of unit electricity consumption, overall furnace efficiency, unit time of melting, and unit furnace productivity). (Author's work.)

follows from the foregoing considerations, the furnace should operate at a current between the current I' and I_1, that is, between the current at the maximum overall furnace efficiency and the current at the maximum arc power. In the former, the minimum electricity consumption is obtained, whereas in the latter, the minimum time of charge meltdown and, therefore, the maximum productivity are achieved.

Knowledge of electric and operating characteristics of an EAF is fundamental for determining the most advantageous electric parameters of the furnace power system. It is decisive to the furnace operating conditions and economics of steelmaking. The actual characteristics can only be made for an existing and operating arc furnace. They are made on the basis of resistance and reactance of the furnace power supply system, whose actual values can only be determined in operating conditions. In addition, the actual value of the heat loss of a furnace can only be established during operation.

4.8 IMPACT OF ELECTRIC ARC FURNACE OPERATION ON THE ELECTRICAL ENERGY QUALITY

In recent decades, the share of energy consumers with nonlinear voltage-current characteristics in the overall power installed in the power system has increased. This resulted from a common use of power electronics drive systems in many industries, including the metallurgical industry. Non-sinusoidal currents drawn from the grid by nonlinear consumers are detrimental to the parameters of the electrical energy used,

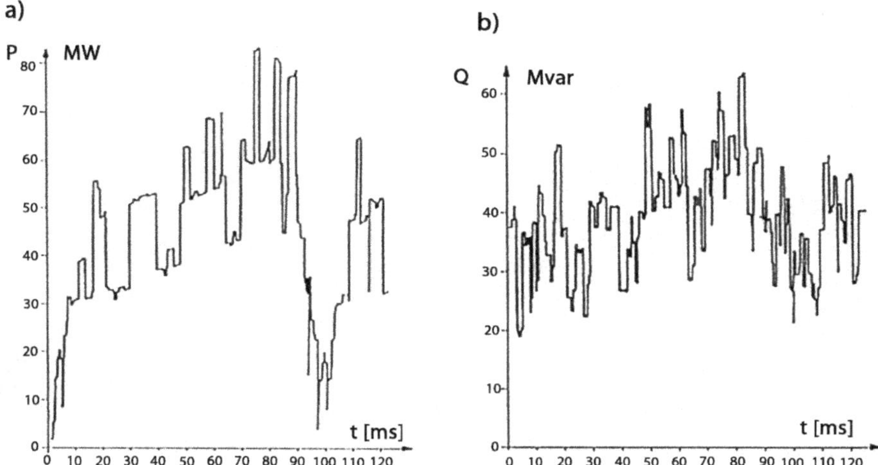

FIGURE 4.21 Examples of active and reactive power changes during a charge meltdown in an electric arc furnace. (Author's work.)

causing voltage drops, which are also non-sinusoidal, which leads to a deformation of the curve of voltage supplied to a customer. It also results in the need to reduce the maximum values of apparent power transmitted by substation step-down transformers. This reduction of power involves the occurrence of additional losses arising from the existence of higher harmonics of current, voltage, and magnetic leakage fluxes. This effect occurs in industrial and municipal grids. The EAF is one of these devices that have the nature of a nonlinear consumer, being a source of higher harmonics generated in the power grid. Examples of active and reactive power changes during charge meltdown in an EAF are shown in Figure 4.21. It can be seen from these curves that fluctuations of both the active and reactive power of around 20 MW, and sometimes more MW (or Mvar, respectively) appear in a few milliseconds. These very fast power fluctuations cause the occurrence of higher-order harmonics in the system, and consequently, the effect of the so-called flickering in the power grid. The most detrimental harmonics are of the order 2, 3, 5, and 7 relative to the power grid [16].

The parameters describing the voltage or current deformation in the grid include the overall voltage deformation factor, expressed in the following equation:

$$THD_U = \frac{\sqrt{\sum_{n=2}^{n=50} U_n^2}}{U_1} 100\% \tag{4.37}$$

For current, this deformation equation has the following form:

$$THD_I = \frac{\sqrt{\sum_{n=2}^{n=50} I_n^2}}{I_1} 100\% \tag{4.38}$$

where:

U_n, I_n – effective values of individual higher-order harmonics of voltage or current, up to the order n,

U_1, I_1 – the effective current value of the first harmonic, of voltage and current, respectively, V, A,

n – the maximum order of harmonic considered according to standards (usually $n=25$, 40 or 50 [17]).

Values of the current deformation factor in power grids THD_I for AC arc furnaces in the charge meltdown stage are from 15% for high-capacity furnace transformers to 35% for low-capacity transformers.

In order to improve the quality of energy in power grids to which the EAF is connected, various solutions to stabilize the arcing are applied in the aspect of reducing its adverse impact. The simplest solution applied is connecting a battery of capacitors with appropriately selected reactive power in parallel to the EAF power system. In modern solutions, electrically expanded compensator systems are applied instead of capacitor batteries, for instance a static var compensator SVC. The structure of a compensator like this consists of capacitors with constant capacitance and reactors with variable inductance, controlled with a thyristor. Therefore, the system reactive power is controlled in real time, depending on the demand arising from optimizing (compensation) operating states of the EAF power system. The compensator is three-phase, and each phase is individually controlled, depending on the state of an appropriate phase of the furnace power system. Eventually, it ensures the balancing of operation of the whole compensation system. Capacitors are divided into sets connected in sections with reactor inductances, which enables individual harmonic components of voltage and current intensity to be more effectively filtered. Compensators of this type reduce the effect of the so-called grid flickering by more than half [18,19].

A device called "Statcom" is another modern solution improving the energy quality in power grids [20]. The device is a kind of voltage converter, which through a set of systems of capacitor and reactor batteries connected in series and in parallel ensures adequate inductive and capacitive reactive power. The device is connected to the grid directly on the primary side of the furnace transformer, in parallel. The basic purpose of the device operation is to reduce the flickering effects in the power grid. The device performance in production conditions presented in publications, for large furnaces indicates that it is possible to achieve a fivefold improvement of the grid flickering indicator. At the same time, the power factor ($\cos\varphi$) is increased practically to a level equal to one. The higher harmonics generated in the system, concerning both voltage and current intensity, are much lower, and the phase symmetry is highly improved.

REFERENCES

1. Smolyarenko V., Popov A. N., Devitaikin A. G., Ovchinnikov S. G., Chernyakhovskii B. P., Egorov A. V.: Next-Generation Electric Arc Furnaces as a Steelmaking Modernization Factor, *Russian Metallurgy (Metally)*, Vol. 2007, 2007, No. 7, pp. 552–559 (in Russian).
2. Publicity materials of General Electric Company.

3. Hubmer R., Dobbeler A.: Improved performance with modern automation solutions for electric steelmaking, *Proceedings of 10th European Electric Steelmaking Conference*, Graz, Austria, 2012.
4. Publicity materials of AEG and Tamini.
5. Publicity materials of Ericable.
6. Publicity materials of Mannesmann Demag.
7. Publicity materials of BRAR Elettromeccanica.
8. Taylor, C. R. (editor): *Electric Furnace Steelmaking*, Association for Iron & Steel Technology, Warrendale, PA, 1985.
9. Picciotto M., Morsut S., Brusa E. G., Bosso N., Zampieri N.: Numerical modeling and dynamic behavior prediction of the AC electric arc furnace structures during the scrap melting process, *Proceedings of 10th European Electric Steelmaking Conference*, Graz, Austria, 2012.
10. Krouchinin A., Sawicki A.: *A Theory of Electrical Arc Heating*, Wydawnictwo Politechniki Czestochowskiej, Czestochowa, 2003.
11. Frittella P., Colombo V., Ghedini E., Solari G.: Application of 3-D time-dependent modeling of plasma arc to an AC electric arc furnace, *Proceedings of 10th European Electric Steelmaking Conference*, Graz, Austria, 2012.
12. Babizki A., Viereck K., Kruger K.: Role of OLTC and its potential in the AC-EAF process, *Proceedings of 10th European Electric Steelmaking Conference*, Graz, Austria, 2012.
13. Kleimt B., Pierre R., Dettmer B., Deng J., Schlinge L., Schliephake H.: Continuous dynamic EAF process control for increased energy and resource efficiency, *Proceedings of 10th European Electric Steelmaking Conference*, Graz, Austria, 2012.
14. Grygorov P., Jepsen O., Theobald F., Hovestädt E., Odenthal H.-J.: Computer modeling and experimental validation of an electric arc furnace, *Proceedings of 10th European Electric Steelmaking Conference*, Graz, Austria, 2012.
15. Karbowniczek M.: *Elektrometalurgia Stali; Cwiczenia, AGH Course Book No. 1364*, Wydawnictwa AGH, Krakow, 1993 (in Polish).
16. Jagiela K., Rak J., Gala M., Kepinski M: Identification of electric power parameters of AC arc furnace low voltage system, *14th International Conference on Harmonics and Quality of Power ICHQP 2010*, IEEE Conference Proceeding, 26–29 September 2010, pp. 1–7.
17. Rak J., Gała M., Jagiela K., Kepinski M.: Analiza obciazenia i strat w transformatorach przeksztaltnikowych ukladow napedowych, Zeszyty Problemowe-Maszyny Elektryczne, No. 89/2011, Katowice, pp. 139–147, 2011 (in Polish).
18. Hackl G., Renner H., Krasnitzer M., Hofbauer C.: Electric arc furnace with static var compensator – Planning and operational experience, *Proceedings of 10th European Electric Steelmaking Conference*, Graz, Austria, 2012.
19. Grunbaum R., Creutzer B., Van Der Rest M.: SVC Brings Productivity Improvements to OneSteel, *Steel Times International*, 2010, No. 9, pp. 234–240.
20. Grunbaum R., Gustafsson T., Hasler J. P., Larsson T., Aigner B., Park D. C.: STATCOM for grid code compliance of a steel plant connection, *19th International Conference on Electricity Distribution*, Vienna, Austria, 21–24 May 2007, Paper 0515.

5 Auxiliary Equipment

Apart from the standard equipment, the electric arc furnace (EAF) is fitted with a number of additional devices to intensify its operation. Heat losses, arising primarily from the heat radiation coming from the external surfaces of the vessel and roof, largely determine the heat consumption in the EAF. The intensity of thermal energy radiation during the melting process is approximately constant over time, and its amount primarily depends on the shell surface temperature. The amount of heat losses or the amount of emitted thermal energy is, therefore, directly related to the tap-to-tap time. In addition, the amount of steel produced in a certain period (e.g. 1 day) directly depends on the tap-to-tap time. To put it simply, it can be said that the tap-to-tap time, on which the heat losses depend, largely determines the steelmaking economics. Therefore, modern EAFs are fitted with additional equipment, which intensifies the melting process, mainly by reducing this period duration. This equipment includes the following:

- oxy-fuel burners,
- lances placed in the so-called manipulator,
- post-combustion lances (burners), and
- porous plugs in the bottom for gas blowing.

5.1 OXY-FUEL BURNERS

A uniform temperature distribution within the whole vessel volume cannot be guaranteed due to the shape of the furnace reaction chamber and the arrangement of electrodes at the ends of which electric arcs – as the source of thermal energy – burn. This causes a non-uniform charge heating and melting rate, and consequently the scrap in the central part of the furnace melts much faster than in the zones that are remote from the arcs, near the furnace walls. In addition, on the wall perimeter, there are areas where less energy from the arcs reaches, which results from the arrangement of the electrodes, and a longer distance from these areas to the burning arcs. Therefore, the so-called hot and cold spots are distinguished within the vessel. To eliminate this effect and to accelerate the spatial heating and melting the charge down, additional heat sources are installed in furnaces in the form of oxy-fuel burners. These burners usually use natural gas, but in some steelmaking shops, fuel oil is used instead of gas [1].

The burners are installed in the area of the "cold spots" present in the vessel. They are deployed in various layouts and numbers in the main door, in the side walls, and in the zone above the taphole.

The application of a single burner placed in the main door is the simplest solution, usually applied in small furnaces, where a single flame from a burner can successfully reach all three underheated areas. The burner is entered through the door, but it is not connected to the other furnace structure. Therefore, the burner can be used

DOI: 10.1201/9781003130949-5

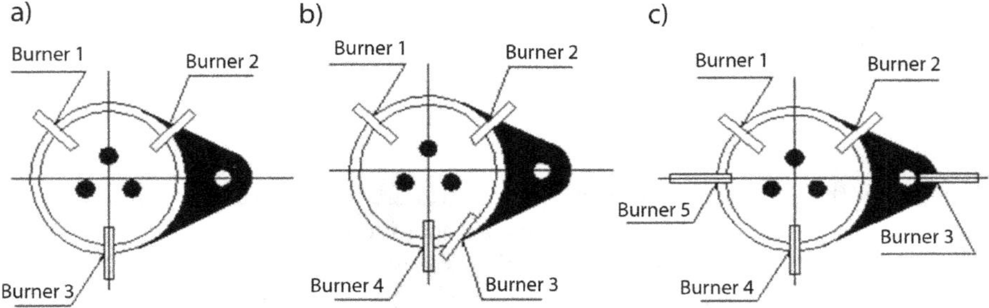

FIGURE 5.1 Possible arrangements of oxy-fuel burners in an electric arc furnace vessel. (a) Layout with three burners. (b) Layout with four burners. (c) Layout with five burners. (Author's work.)

in a very flexible manner, subject to the process needs. In the melting periods when the burner operation is not needed, it is retracted from the furnace, which contributes to lower gas and oxygen consumption and increases its life. The location of the burner in the door has its advantages: the heat stream can be directed toward a specific area, and the amount of generated off-gases is relatively smaller. At the same time, the burner positioning during melting needs to be accurately controlled, which requires high operating qualifications, but it enables good performance indicators to be achieved. The burner is placed in a special manipulator situated in front of the door. The manipulator enables the operator to precisely and quickly control the position of the flame within the furnace vessel – guiding it to the desired area.

In medium- and large-sized furnaces, where the vessel dimensions are much larger, arrays of three, four, five, and sometimes more burners are applied. Various systems of burner arrangement within the vessel of the furnace are shown in Figure 5.1a–c.

In the case shown in Figure 5.1a, three burners are installed in the furnace wall, located in the zones between the electrodes, that is, in the cold spots. For the layout shown in Figure 5.1b, the fourth burner is installed, apart from the three burners arranged as in case (a), and its flame is directed toward the balcony part area with the taphole. The objective of this solution is to reheat the balcony, where the distance to the arc flame is long and less thermal energy coming from the arcs reaches this area. The layout shown in Figure 5.1c also contains three burners, as in the previous solutions, but it also has two burners: one with the flame directed toward the balcony area and the other placed in the main door. The above-mentioned solutions are shown as examples. In practice, other solutions of the oxy-fuel burner layout are also possible, adapted to individual practice needs of a specific steelmaking shop.

The method of mounting a burner in the wall, in the vertical plane, should ensure the optimum conditions for the penetration of the flame toward the desired areas, or the "cold spot". Such areas exist near the furnace walls between the electrodes. It is assumed that the angle between the burner position line and the horizontal plane should be from 20° to 45°, which is shown in Figure 5.2. Such burners are permanently mounted on the furnace wall and they are immovable. Burner no. 3 in Figure 5.1c is fixed in the top roof part of the balcony zone, with its flame directed toward the taphole area.

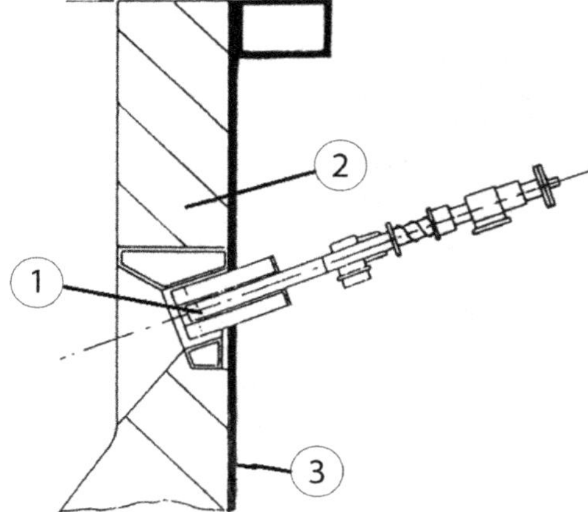

FIGURE 5.2 Position of the oxy-fuel burners in the vertical plane of the electric arc furnace wall: 1 – oxy-fuel burner, 2 – vessel wall, 3 – vessel shell. (Author's work.)

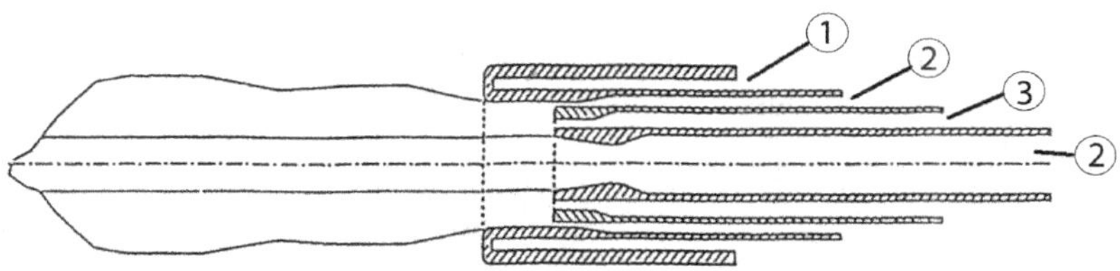

FIGURE 5.3 General diagram of an oxy-fuel burner section: 1 – water, 2 – oxygen, 3 – gas. (Author's work.)

The selection of the burner arrangement in the furnace also impacts its operating conditions. For each furnace, this layout should be designed individually. The following parameters influencing the optimal utilization of the burners should be considered:

- furnace vessel dimensions and the hearth shape,
- average density of the scrap applied,
- capacity of the baghouse,
- desired increase in production,
- frequency of burner use.

A classic oxy-fuel burner is designed as four coaxial pipes. See Figure 5.3 for a general diagram of a burner section.

The inner pipe is for the oxygen supply. The fuel gas supply pipe is laid on it. The next pipe laid on the gas one is used to supply oxygen. The last one (the outer one) is a pipe leading the cooling water. The purpose of supplying oxygen with two pipes is to improve mixing oxygen and fuel before burning. The outer pipe supplies the cooling water to the whole burner structure. The tips of the pipes supplying oxygen and gas have variable diameters and are adjusted to the requirements of the de Laval nozzle. The water-cooling pipe is longer. A design like this enables the fuel combustion process to be optimized, and it protects the burner from overheating.

The outer pipe of the burner is made of stainless steel and has a copper tip. Its purpose is to minimize slag adhesion and to protect from the arc thermal energy. If natural gas is used as the fuel, its pressure is 0.35 MPa, the oxygen pressure is 0.8 MPa, and the cooling water consumption is about 6 m^3/h, at a pressure of 0.35 MPa. The burner operation is controlled from the operator's panel. In industrial practice, in early burners the problem of burner wall burn-through occurred, and the burner face had to be protected against nozzle clogging. The tip is clogged as a result of slag-metal liquid bath splashes flowing down from the furnace walls onto the burner as well as due to slag "spattering" onto the burner face. The intensity of spattering depends on the metal charge quality, burner installation level, their angle of inclination, etc. This problem is aggravated, in particular, when operating EAFs without cooling panels. That has been solved, however, by applying protective shields above the burners or using a graphite protective burner tip and an additional water-cooled copper head.

Modern oxygen lance heads enable oxygen to be injected at velocities above the speed of sound (so-called supersonic lances). Therefore, the oxygen jet reaches the metal bath with a higher kinetic energy, which improves its penetration into the melt and contributes to the intensification of the stirring processes and oxidation reactions. The burner can also be ended with a tip, which was designed to create two wide flames directed toward the walls. Such burners feature two tips with a 70° angle between the flames, which are inclined at the 20° angle to the level. A burner with a tip of this type uses 6–7 Nm^3 gas and 14–15 Nm^3 oxygen per metric ton of steel manufactured [2].

To ensure a constant level of oxygen and constant gas pressures during the burner operation, the sections of the system supplying the mentioned gases to the burner should be appropriately designed. When considering the oxy-fuel burner system, one should pay special attention to its safety operation. In order to avoid the possibility of creating an explosive mixture in the furnace reaction chamber, properly mounted safety valves are used in the system. A well-designed burner system should have valves for low- and high-pressure oxygen, low- and high-pressure natural gas, water, and ignition.

The number of burners and their capacity is adjusted to the furnace capacity, vessel geometry, and the type of scrap applied. The capacities of the burners, for large furnaces, are as high as a few megawatts.

One of the most important parameters of the burner operation is the ratio of oxygen to natural gas, determining the gas combustion effectiveness and the transfer efficiency of the combustion process chemical energy to the scrap that is heated. The combustion of gas in oxygen can be complete, which means that the oxygen share per one mole of gas is selected so as to obtain only one component in the combustion products (flue gas) – the product of the combustion reaction, and zero gas and oxygen. Such combustion is called stoichiometric combustion because the stoichiometry is maintained in the chemical reaction of combustion. For the combustion of natural gas (containing approximately 80% methane CH_4), the oxygen share is 2 moles per 1 mole of natural gas. In practice, three possible cases of the oxygen to natural gas ratio are applied (in the burner operation, in the EAF conditions):

- So-called lean (substoichiometric), meaning the ratio of oxygen to natural gas is less than 2:1; then the so-called deoxidizing (reducing) flame is obtained, recommended for the production of stainless or highly alloyed steel grades, where the oxidation of the metal bath constituents should be avoided. The disadvantage of such a ratio is a lower flame temperature and a reduced effectiveness of burner operation.
- Stoichiometric, meaning the ratio of oxygen to gas equals 2:1; then the normal flame is obtained, the most effective in terms of energy. The flame is characterized by the highest temperature and the highest efficiency of transfer of energy to the scrap; it is preferred for an EAF due to the insignificant oxidation of carbon and metal bath constituents.
- So-called rich (superstoichiometric), meaning the ratio of oxygen to gas is higher than 2:1; then the flame containing oxygen is obtained. Admittedly, the energy efficiency is lower, but the so-called lance effect occurs, or oxygen in the flame, which causes the additional oxidation of metal bath constituents. Depending on the charge type, it is possible to operate with excessive oxygen for a short time, for instance at the beginning of the burner operation cycle in order to increase the flame penetration; however, this operation is not always economically viable.

In the classic steel production process, for most process operations, the stoichiometric variant is the preferred ratio of oxygen to natural gas. The operation of oxy-fuel burners in an EAF is the most effective at the initial phase of the steelmaking process, during scrap, when the scrap temperature is relatively low. As the scrap temperature increases, the burners become less efficient. Figure 5.4 is a graph showing the burner efficiency during the charge melting.

The efficiency as regards oxy-fuel burners in an EAF means the ratio of the heat amount that is absorbed by the metal bath to the heat amount that is theoretically emitted from the gas combustion. As can be seen from the curve in the graph, the

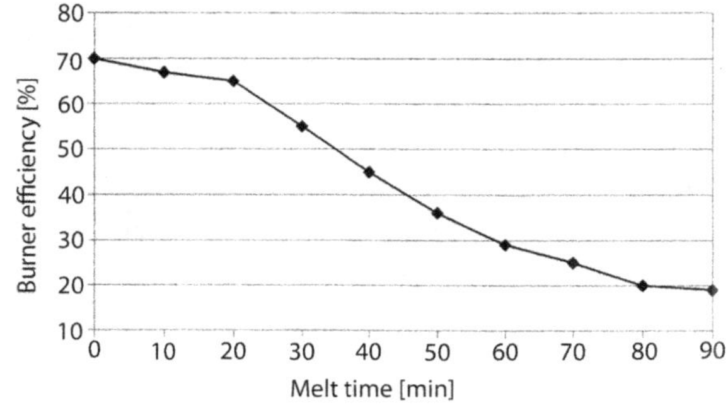

FIGURE 5.4 Burner efficiency during the charge melting. (Author's work.)

efficiency of the burner energy transfer is a function of the scrap melting time. The burner efficiency exceeds 60% for one-fifth of the melting time, and by the halfway mark of the melting time, it drops to approximately 35%. At the end of the charge melting period, it reaches approximately 20%. Usually, the off-gas temperature increase determines the time when the use of the burner should be stopped. When the efficiency falls below a set value, the burners are switched to standby, that is, the operation at minimum power, just to keep the burner flame going.

Examples of burner operation times during charge melting, for two variants of loading baskets:

- *For a charge loaded with three baskets*: two-third of the first basket meltdown time, half of the second basket meltdown time, and one-third of the third basket meltdown time.
- *For a charge loaded with two baskets*: half of the first basket meltdown time and half of the second basket meltdown time.

As a rule, the practical application of oxy-fuel burners enables approximately 25% of electrical energy needed for the manufacturing process to be replaced with chemical energy from gas combustion in the burners. The share of this energy can vary depending on the furnace characteristics, the range of steel grades manufactured, and the technological practices applied. The level of capacity at which the burners operate is limited by the rate at which the scrap can absorb heat generated by the burners. The optimum burner capacity is achieved when the effectiveness of heat transfer to the scrap is at its maximum. The desired time of burner operation during charge meltdown can be determined on the basis of experience in the operation of a specific furnace and process empirical models.

The application of oxy-fuel burners in the EAF operation causes an increase in the volume of off-gases forming in the charge meltdown process. This volume may increase by approximately 50% compared to the classic process without burners. At the same time, the amount of thermal energy and the temperature of these gases increase. It contributes to an increased so-called thermal load of the roof and the off-gas collection system operation.

The application of oxy-fuel burners as an additional source of thermal energy in the EAF contributes to a reduction of electricity consumption and the shortening of the tap-to-tap time. Depending on the furnace size, the burner capacity, and the practices applied, it is possible to obtain at least a 10% reduction of electricity consumption as well as a reduction of the tap-to-tap time by 15%–30%. As a result, the consumption of graphite electrodes also decreases (the electrode consumption can be reduced by up to about 12%). At the same time, the burners use gas and oxygen. The demand for natural gas, depending on the burner capacity, is from 5 to 20 m^3 of gas/1 [t] of steel produced. In some cases, other fuels are used instead of natural gas, for instance, coke-oven gas or heavy oil (mazut). The demand for oxygen for the combustion of the fuel in the burners depends on the type of fuel and the assumed excess oxygen ratio; this demand is from 10 to 30 m^3/[t] of steel produced. An increase in the charge melting rate in a furnace fitted with burners increases the life of the bottom refractory lining (a shorter time of contact with the melt). However, due to the application of burners, the designed and installed dust removal system must have a higher capacity to discharge an increased amount of off-gases, containing more dust and featuring higher temperatures.

The burners enable the productivity to increase with a lower consumption of electrical energy and electrodes. Additional advantages from the application of burners include a better furnace operating flexibility, arising from the ability to control the charge melting rate, and thereby the tap-to-tap time, which streamlines the operation in the entire steelmaking shop system: EAF – ladle furnace – continuous casting. The correct arrangement of burners contributes to a more uniform charge heating and melting within the furnace vessel volume, thereby the operating conditions of all furnace structural components improve and their life extends.

5.2 MANIPULATOR

A manipulator is a device that enables the lances for blowing various media into the furnace vessel to be installed. It consists of a vertical mast made of a steel structure and a head, installed in the upper part of it, supporting the lances and allowing them to move. The mast is installed in front of the main door of the furnace in two ways. It can be permanently fixed to the service platform or it can be placed on a carriage, which enables it to be moved, even during the lance operation. The design of mounting the lances to the mast head must enable their tips to move so that the materials injected by the lances can penetrate anywhere within the furnace vessel volume. The lance operation, including the optimization of the amount of the material fed, time of their operation, and position of their tips within the furnace vessel volume, is controlled from a panel located in the operator's cabin.

Usually two or three lances are installed at the manipulator for injecting oxygen gas, coal materials for slag foaming, or slag-forming materials. In practice, two types of lance design are applied. In the first solution, the lance is made in the form of a single, regular steel pipe, with a diameter from 20 to 50 mm. A lance like this is consumed during the operation (it gradually dissolves in the metal bath). In the second case, the lance is made as a system of a few concentric steel pipes, ended with a special head containing a nozzle taking the material outside; the design of such lances also enables them to be water-cooled. Lances of this type are not dissolved in the melt. Water-cooling ensures that the lower temperatures of the lance tip (including its head) are maintained. Figure 5.5 shows the layout of the manipulator in relation to the furnace.

Figure 5.6 shows an example of the lance layout placed on the manipulator during operation. In this case, the top lance injects fine lime to the slag volume in order to increase its basicity. The lance in the middle injects fine coal material into the boundary area between the melt and slag in order to foam the slag. The bottom lance injects oxygen gas to the melt volume to oxidize the metal bath constituents. The time of operation of individual lances depends on the steelmaking practice in the furnace. Such a method of material feeding (oxygen gas or solid coal or lime particles) to the metal bath or slag ensures the most advantageous conditions for the utilization of these materials. Both oxygen and solid particles injected with this method are used for the operation of appropriate metallurgical processes to the utmost compared to

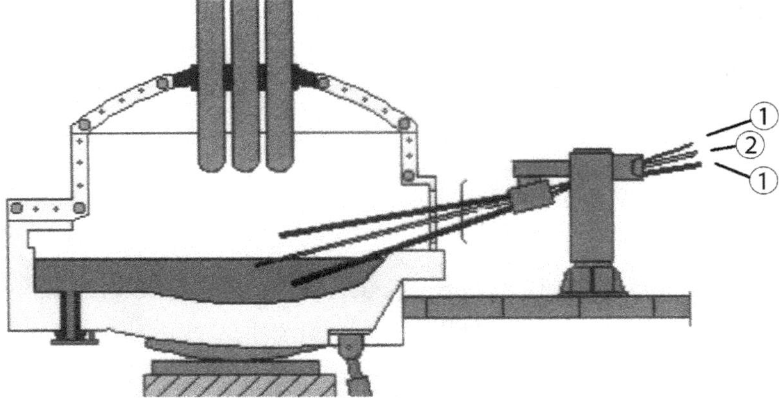

FIGURE 5.5 Diagram of the installation of the manipulator with lances in relation to the electric arc furnace: 1 – oxygen, 2 – coal. (Author's work.)

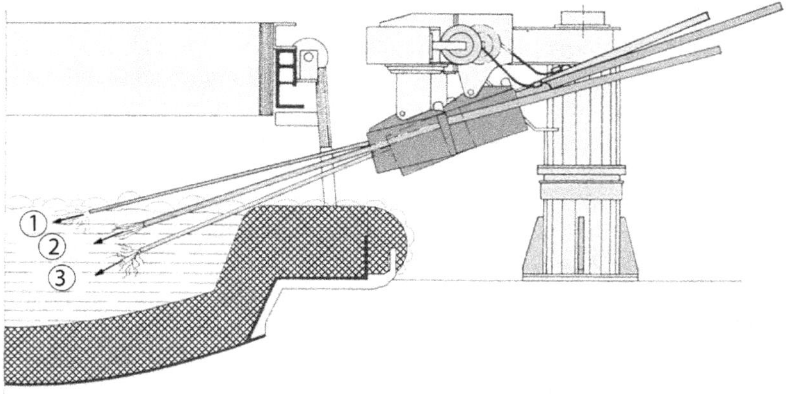

FIGURE 5.6 Diagram of the lance position on the manipulator during the injection of oxygen, coal, and lime: 1 – lime, 2 – coal, 3 – oxygen. (Author's work.)

the other methods of their addition. The amounts of these materials not taking part in the processes (losses) are negligible.

One of the solutions of lance application in the EAF is a design enabling two different materials to be injected at the same time. In this case, the lance construction involves a system of a few concentric pipes, with a head with two appropriately arranged nozzles at their ends [3]. This solution has the advantage of easier operating control of a single lance injecting two materials simultaneously. However, there is a disadvantage, which is the more complicated operation from the perspective of supplying two different materials to the desired area of the furnace vessel volume. Another disadvantage is less flexibility, if the individual amounts of materials fed need to be changed frequently during the steelmaking process in a different manner.

Another solution for the simultaneous injection of oxygen and coal into the furnace vessel is the installation of a special, water-cooled structure in the furnace wall, where two lances are placed: one for oxygen and the other for coal. An example of this type of design is JetBox™ [3].

5.3 POST-COMBUSTION LANCES

During the steelmaking process in an EAF, a substantial amount of gaseous carbon oxide is emitted. This gas mainly forms in the volume of a metal bath or slag as bubbles and floats to the reaction chamber. In the classic case, the gas is sucked into the gas exhaust system together with other gases, where it is burned to CO_2. There are also solutions to enable CO to be burned to CO_2 within the furnace reaction chamber. The post-combustion reaction is exothermic and involves emitting considerable amounts of thermal energy. Thereby, the utilization of this energy inside the furnace vessel increases the energy efficiency of the melting process. In fact, it is not possible to fully utilize this energy for the manufacturing process due to the location of the reaction occurrence, but part of it can be recovered. The post-combustion reaction occurs above the melt and slag level, and the emitted thermal energy primarily heats the top part of the walls and roof. The metal bath is heated with this heat to a lesser degree. Thereby, the post-combustion of CO contributes to an increased thermal load of the roof and furnace elbow. Nevertheless, the application of post-combustion brings energy benefits in the form of less heat losses, in particular in the top part of the vessel.

Usually, oxygen lances are the devices performing the post-combustion of the carbon oxide emitted in the process to CO_2. They can be placed at the manipulator or installed permanently in the furnace wall. Then, the lance is made as a regular, single steel pipe, with a diameter in the range of 20–50 mm. Oxygen gas is fed through the lance. Solutions where oxy-fuel burners are designed to enable them to operate as both a burner and a post-combustion lance are also applied. Then, the devices operate in the initial charge melting period as burners, while in the final period as post-combustion lances. Usually, the burners of this type are characterized by a lower thermal power (one burner has a power of 2–6 MW). A system of oxygen lance called Oxiarc for the post-combustion of carbon oxide in EAFs [4], patented by Messer, is an example of such a solution.

To determine the optimal demand for oxygen supplied with the lance, it is necessary to know the exact chemical composition of off-gases at the furnace vessel outlet. To this end, special probes taking gas samples are applied to enable hot off-gases containing a large content of dust to be continuously analyzed. Therefore, it is possible to adjust the proper amount of oxygen for post-combustion, to the current composition of the off-gas (mainly the CO content).

The Oxiarc-EAF lances are mounted on the upper part of the water-cooled panel of furnace vessel, in the area of the elbow connection in the off-gas line so that the oxygen flows contrary to the direction of the off-gas. A jet of oxygen is introduced downward at an angle of 30° to the level and tangentially to the space between the electrode circle diameter and the furnace wall. A strictly defined amount of oxygen is fed by the control system. The added amount correlates to the electrical load of the furnace. It ensures that oxygen continuously mixes with the off-gas and results in almost complete combustion.

The practical savings of electrical energy arising from the application of selective post-combustion in oxygen are from 2.5 to 4.7 kWh per cubic meter of oxygen. To ensure the good mixing of gases with the injected post-combustion oxygen within the furnace volume, it is necessary to ensure sufficient free space and to ensure that the oxygen jet does not reach the surface of water-cooled panels constituting the furnace walls. It is thereby necessary to properly position the lance in relation to the walls and to adjust the parameters of the oxygen that is injected to the size and shape of the furnace [5–11].

REFERENCES

1. Toulouevski Y. N., Zinurov I. Y.: *Innovation in Electric Arc Furnaces*, Springer, Berlin, 2010.
2. Bergman K., Gottardi R.: Design Criteria for the Modern UHP Electric Arc Furnace with Auxiliaries, *Ironmaking and Steelmaking*, Vol. 17, 1990, No. 4, pp. 156–159.
3. Publicity materials of Air Products PLC.
4. Publicity materials of Messer.
5. Krasssnig H., Kleimt B., Voj L., Antrekowitsch H.: EAF Post-Combustion Control by On-Line Laser-Based Off-Gas Measurement, *Archives of Metallurgy and Materials*, Vol. 53, 2008, No. 2, pp. 455–462.
6. Millman M., Nyssen P., Mathy C., Tolazzi D., Londero L., Candusso C., Baumert J. C., Brimmeyer M., Gualtieri D., Rigoni D.: Direct Observation of the Melting Process in an EAF with a Closed Slag Door, *Archives of Metallurgy and Materials*, Vol. 53, 2008, No. 2, pp. 463–468.
7. Mathy C., Nyssen P., Brimmeyer M., Gualtieri D., Rigoni D., Baumert J. C.: Innovative Technique for Reliable Operations and Blow-Back Prevention of EAF Annular Burners, Combined Burners and Injectors, *Archives of Metallurgy and Materials*, Vol. 53, 2008, No. 2, pp. 469–473.
8. Brhel J., Shver V., Farmer C., Novák M., Heide R., Domovec M., Mastelák M., Kucera J., Tlamicha P.: The Latest Experience with Advanced Chemical Energy Introduction to Smaller Size Furnaces, *Archives of Metallurgy and Materials*, Vol. 53, 2008, No. 2, pp. 489–493.
9. Gareth M.: SVC Brings Productivity Improvements to OneSteel, *Steel Times International*, Vol. 3, 1992, pp. 14–16.
10. Jones J. A. T.: New Steel Melting Technologies – Oxy – Fuel Burner Application in the EAF, *Iron and Steelmaking*, Vol. 23, 1996, No. 5, pp. 63–65.
11. von Scheele J., Selin R., Orrebo K. A., Palmgren J., Moder R.: Industrial Gas Applications – A Key to Improved Iron and Steel Making, *Berg und Huttenmannische Monatshefte*, Vol. 143, 1998, No. 5, pp. 189–194.

6 Graphite Electrodes

Graphite electrodes are the final section of the arc furnace power supply system, and they are used to supply electricity to the vessel's reaction chamber. Electric arcs burn at the tip of the electrodes, where the supplied electric energy is transformed into heat energy that is used for the metallurgical process. Both the specific consumption of electricity and the specific consumption of electrodes have a significant impact on the economic efficiency of steel production in electric arc furnaces (EAFs). At the same time, these values can be easily and accurately determined on the basis of the total electricity consumption and the electrodes consumed to produce a certain amount of steel. These indicators depend on the furnace design parameters, the quality of the electrodes, and a number of engineering factors. Therefore, the proper selection of the electrode quality and the favorable working conditions determine the costs of steel production [1–5].

6.1 PRODUCTION OF ELECTRODES AND THEIR PROPERTIES

Electrodes are produced from special grades of anthracite, petroleum and pitch coke, electrode scrap, and graphite. Tar and anthracene oil are used as binding agents. The shares of particular materials are individually selected in accordance with the patent rights of manufacturing companies, and they determine the manufacturing costs and quality of the electrodes produced.

Anthracite is a type of coal with the highest degree of carbonization of vegetable substances. The structure and content of volatile matter in anthracite depend on its degree of carbonization. Older anthracite is best suited for the production of electrodes due to its low volatile matter content, cleavage difficulty and high density, mechanical strength, and thermal resistance. Anthracite is the basic raw material used for the production of electrodes. By thermal refining at temperatures of 2,073–2,673 K (1,800–2,400°C), anthracite can be used to obtain electrographite with a high degree of purity (approximately 0.2% ash) and a well-developed crystalline structure similar to that of flake graphite.

Due to the ease of graphitization, petroleum coke is the most suitable raw material for the production of electrodes. It is obtained by the heat treatment of residues from petroleum distillation. From the point of view of the quality of electrodes produced, it is important that the petroleum coke used in their production has a needle-like crystal structure and not a lamellar one. Petroleum coke that is used for the production of electrodes should have the lowest possible content of volatile organic compounds and ash, and its sulfur content should never exceed 1%.

Pitch coke is obtained by coking hardened pitch obtained from bituminous coal in the production of metallurgical coke or coke for the power industry. The resulting pitch coke has a crystalline graphite structure. Thanks to its high purity, good

DOI: 10.1201/9781003130949-6

mechanical strength, and low electrical resistance, it can be used in certain quantities as a substitute for expensive and scarce petroleum coke.

Natural graphite comes in two basic crystalline varieties, namely, as coarse crystalline graphite (scale, flake) in the form of large, thick scales with a metallic sheen and as fine crystalline graphite in the form of fine crystals or dust with a weak sheen. The fine crystalline graphite is used for the production of electrodes. Due to its structure, natural graphite features a high electrical conductivity, which is particularly important in terms of the function of electrodes in an EAF.

In order to impart adequate plasticity to the solid carbon raw materials used in the production of electrodes, which is necessary for their formation, the grains of these materials are moistened with a binder. Materials with a high carbon content, low ash content, and low viscosity within a specific temperature range are used as binders. Pitch made of bituminous coal tar meets these requirements very well. It is relatively cheap, easily accessible, and contains up to 93% carbon. In order to lower the pitch viscosity at the temperatures occurring when the mass is mixed and the electrodes are formed, anthracene oil is added to it.

These materials are first subjected to the preliminary crushing, calcining, shredding, grinding, and sorting. For the production of electrodes with a diameter of up to 300 mm, the materials must be shredded to a grain diameter of maximum 4 mm. For electrodes with a larger diameter, the acceptable grain size is 10 mm. From the material prepared this way, a feedstock mix is made, in which shares of individual grain fractions are strictly selected. It must be made sure that the share of the subgrain fraction below 0.3 mm is about 40%. This ensures the good filling of empty spaces between grains of larger fractions. The feedstock is then subjected to stirring, granulometric homogenization, and moistening with a medium-temperature pitch, in the amount of about 2%. After the materials are mixed, an electrode mass is obtained, from which electrode sections are formed in steam-heated presses. The semi-finished product obtained after pressing is baked in an inert or reducing atmosphere. The process takes about 2 weeks, according to the so-called temperature baking curve, which determines the heating rate, holding time at the assumed temperature, and cooling rate.

During this process, the quality characteristics and operational properties of the final product (electrode) are primarily formed. Baking is carried out in electric or gas-fired heating furnaces in which the electrode produced is covered with a protective powder consisting of coke breeze with various supplementary additives. During baking, a coke mesh is formed from the binder, which gives the mechanical strength and other necessary physical properties to the electrodes. The baking process is carried out at temperatures of up to 1,373 K (1,100°C), with different heating rates of 1.2–3.5°C/h in the zone of slow heating, and the baking time is 300–400 h.

In order to further improve the functional parameters, the baked electrodes are graphitized. The purpose of the graphitization process is to transform the carbon contained in the materials used for the production of electrodes into a crystalline form of graphite. The material obtained by graphitization is polycrystalline and has properties somewhat similar to those of single graphite crystals. The graphitization is carried out in Acheson type electric resistance furnaces characterized by a structure where the electrodes that are graphitized are heating elements, and the space between the electrodes is filled with a protective coal powder. Thanks to this resistance furnace design, a reducing atmosphere around the electrodes protects them from oxidation, and the obtained temperature of the electrodes may exceed 3,273 K (3,000°C). The graphitization process takes 50–70 h at a temperature of 2,773–3,073 K (2,500–2800°C). The total time of the graphitized electrode production cycle is approximately 60 days. The specific electricity consumption is 3,000–5,000 kWh/t of graphite products.

The electrodes should have the following properties:

- high electrical conductivity,
- high resistance to high temperature,
- small thermal conductivity,
- low consumption per steel production unit,
- sufficient mechanical strength.

In terms of the capability to carry electrical loads, electrodes are classified as follows:

- standard power electrodes, marked with RP (regular power),
- high power electrodes, marked with HP (high power),
- saturated and impregnated, marked with SHP (super high power),
- very high power, marked with UHP (ultra-high power).

The basic properties of graphite electrodes are given in Table 6.1.

Electrodes subjected to the classic graphitization process achieve a porosity of 30%, which significantly limits their functional properties. In order to obtain products with better performance parameters, the graphitized electrodes are saturated or coated with special materials afterward. Usually, a mixture of medium-temperature pitch and anthracene oil or prepared tar is used for the saturation. After saturating the electrodes, it is necessary to apply a thermal treatment. The saturation of electrodes reduces their porosity and specific resistance, which in turn reduces their oxidizability and allows the acceptable current density to be increased.

In some cases, electrodes are coated to increase their surface electrical conductivity and reduce oxidizability. This can be done by the metallization of the side surface by spraying with a plasma torch. The practice of such coating is relatively simple. The electrode, after machining, is given an appropriate surface roughness with metal

TABLE 6.1
Basic Properties of Graphite Electrodes (Author's Work)

Parameter	RP	HP	SHP	UHP
Apparent density (g/cm³)	1.55–1.60	1.60–1.65	1.66–1.69	1.70–1.75
Specific resistance (μΩm)	8.0–9.5	6.0–7.0	5.9–6.3	4.5–6.0
Flexural strength (MPa)	8.0–10.0	10.0–12.0	11.5–13.0	12.5–15.0
Thermal expansion coefficient ($1/K \cdot 10^{-7}$)	13–15	15–17	14–16	4–10
Ash content (%)	Max. 0.30	Max. 0.20	Max. 0.15	Max. 0.10

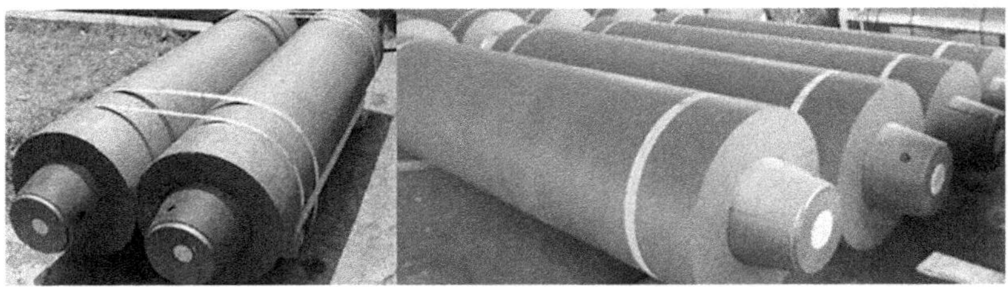

FIGURE 6.1 Graphite electrode sections with screwed-in biconical connectors (pin or the so-called nipple joining system). (Author's photos.)

brushes, and then a thin layer of metal (e.g., aluminum) is applied on it with a plasma torch. The second plasma torch evens out the thickness of the metal layer, and by blowing appropriate oxides, it is given a crystalline structure, which ensures good electrical conductivity and low oxidizability.

After graphitization, each electrode section is machined by turning its side surfaces, cutting and leveling its front faces, and drilling and threading connection sockets. Graphite electrodes are produced in sections with standardized diameters, expressed in inches: 10, 12, 14, 16, 18, 20, 22, and 24. The lengths of the sections are also standardized: 1,500, 1,800, 2,100, 2,400, and 2,800 mm. Figure 6.1 offers a view of the electrode sections.

The electrodes can be connected by means of cylindrical (used for small diameter electrodes) or biconical connecting pins (for larger diameter electrodes). The production method of connecting pins is the same as for sections, and the quality and property requirements must also be the same. The dimensions of connecting pins are also standardized. Their diameters, depending on the electrode diameter, range from 200 to 320 mm and their lengths from 200 to 450 mm. By connecting individual electrode sections, an electrode column of the required length is obtained. The method of connecting electrodes is shown in Figure 6.2.

6.2 ELECTRODE CONSUMPTION

The wear of graphite electrodes under the operating conditions in an EAF is an extremely complex process. It consists of a high temperature impact, the effect of an electric arc, the chemical impact of the metal bath, slag and hot oxidizing gases, and many diverse mechanical loads [6]. The general diagram of the consumption of electrodes is presented in Figure 6.3. The total consumption of electrodes is divided here into process-related and accidental.

The process-related consumption depends on the specific operating conditions of the arc furnace. For a given type and quality of electrodes, using a specific practice and manufacturing specific steel grades, for a given furnace, it only depends on the operating conditions of the furnace. In the electrode tip area, the fast moving arc, liquid slag, and metal bath cause gradual erosion of the graphite. Similarly, the gradual taper of the side surfaces toward the electrode tip is caused by the oxidizing action of the furnace atmosphere. The linear shortening of the electrode length is actually only achieved during arcing (i.e., when power is supplied to the furnace), while the

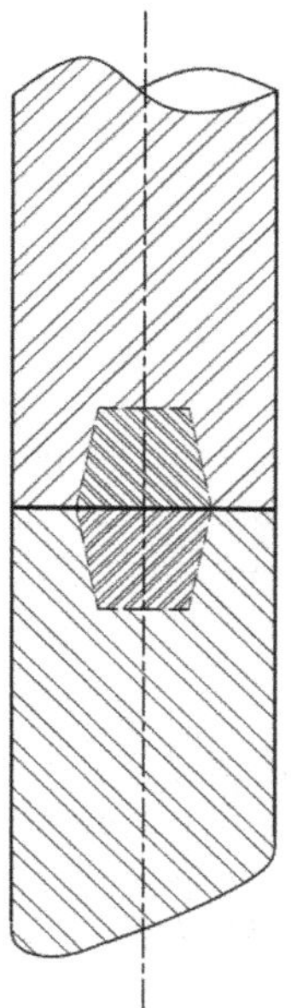

FIGURE 6.2 Diagram of the method of connecting sections of graphite electrodes. (Author's work.)

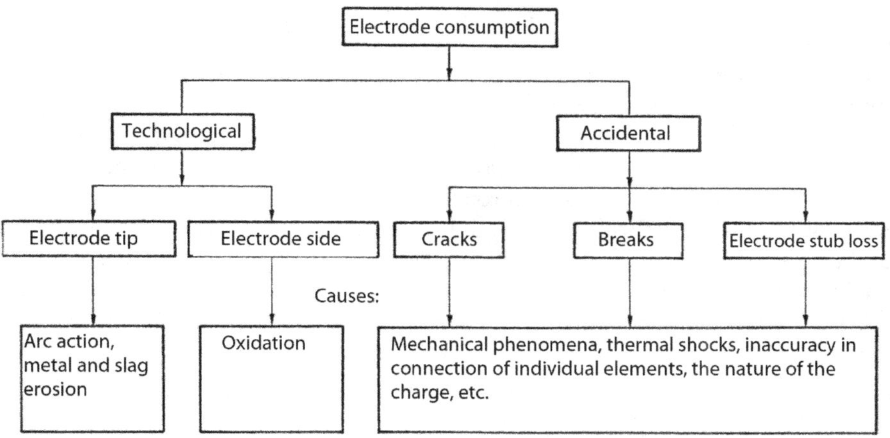

FIGURE 6.3 Breakdown of graphite electrode consumption. (Author's work.)

side surface wears practically all the time when the electrode is exposed to a high temperature. The combination of these two factors determines the so-called process-related consumption of electrodes [7].

Regardless of the gradual wear of electrodes, their length can suddenly shorten as a result of breakage, butt tear-off, or splitting. This can be caused by poor scrap preparation in the charging basket, malfunctioning of the automatic electrode positioning system, or the incorrect connection of sections in the electrode column.

Factors influencing the consumption of electrodes in EAFs can be divided as follows [8]:

- chemical factors, which include the oxidizing effect of the furnace working atmosphere and the reactions of the metal bath and slag with the electrode,
- mechanical factors such as impacts, electrodynamic forces, and furnace tilt,
- thermal factors such as the sublimation of the electrode tip in the arc zone and sudden temperature changes.

6.2.1 Wear of the Electrode Tip

The linear wear of the electrode tip is caused by the impact of the electric arc, oxidizing atmosphere as well as erosion and chemical reactions with the metal bath and slag. The reasons for the occurrence of the above factors depend on the specificity of the electrode operation in the various stages of the steelmaking process.

The period of charge meltdown is the most difficult in the operation of electrodes. During this time, the electric arc is burning between the electrode and the scrap. However, for the arc to strike, the electrode tip must touch a solid piece of the metal charge. Then, the intense point heating of the electrode occurs, and after it has been detached from the metal charge, electrons are emitted from the heated spot. For such a phenomenon to occur, a current exceeding the rated current of the furnace transformer by 1.5–2 times must flow. Such high currents, although very short in duration, pose a serious hazard to the entire electrical system of the arc furnace, including the electrodes. Accidental short circuits with caving scrap are even more dangerous for the electrical system, and particularly for the electrodes, because their duration is usually much longer. Often, during the charge meltdown, and especially during electrical short circuits, the current periodically flows at one point of the electrode for a longer time, not being able to rotate on its tip face. Then, a relatively stable, permanent, long arc is formed, clearly damaging the electrode. There is no sufficient explanation or description of the phenomena taking place at that time, but it is only known from the operational practice that the electrode tips crack during this period [6]. It is believed that this may result from both electrical short circuits and long burning of the permanent arc at one location. These cracks not only cause individual pieces of electrodes to fall off, but in special cases, it may also cause the last section to fall off from the electrode column.

The described consumption of electrodes occurs largely in the operation of very high power arc furnaces. The causes and mechanism of significant wear of the electrode tip as a result of cracking are related to their relatively poor heating during the meltdown period, local arc burning, and its very variable length [9]. In addition, electrical short circuits and huge local current concentrations often occur. This causes significant temperature gradients at the tip of the electrode, thereby resulting in cracks.

The wear of the electrode tip during the furnace operation after the charge meltdown is smaller, but still significant. The current density at the arcing locations is still high, but the arc is then able to rotate, although its impact zones on the electrode face are limited. This is due to the impact that is exerted on the arc by the forces from the electric fields of the two arcs burning at the tips of the other electrodes. Therefore, the arc only operates basically at the outer face of the electrode tip. Consequently, the electrode tip wears asymmetrically, which depends on the arc length. Long arcs cause a greater difference in the consumption on the outer and inner part of the electrode tip face than short arcs. Very short arcs can even form a concave electrode tip [9].

The distribution of the current density on the electrode cross section is uneven, which is caused by the skin effect, mutual induction, and the variability of the specific resistance depending on the temperature and different thermal conductivity in the different parts of the electrode and joints. At about 250 mm from the electrode tip face, the current begins to concentrate toward the arc root. The temperature of the root is 3,573–3973 K (3,300–3,700°C), while in the rest of the tip it is only 1,973–2,273 K (1,700–2,000°C). The diameter of the arc root ranges from 6 to 13 mm [9]. Therefore, the most intense linear consumption of the electrode tip occurs at the arc root zone. Large temperature gradients cause the phenomena of thermal expansion in the arc root zone, which causes the formation of radial and axial forces that can contribute to the loss of graphite particles. The high temperature also allows the graphite to sublimate from places directly adjacent to the arc root. In the periods of operation of the furnace following the charge meltdown, when the currents are already lower or the arc does not burn at all, the destruction of the frontal part of the electrode tip also occurs. This stems from chemical reactions between the graphite and the liquid slag, and the metal bath. The chemical reactions taking place at this time can probably be written as follows [8]:

$$(\mathrm{FeO}) + C_{gr\ \text{from electrode}} = [\mathrm{Fe}] + \{\mathrm{CO}\} \tag{6.1}$$

$$(\mathrm{CaO}) + 3C_{gr\ \text{from electrode}} = (\mathrm{CaC_2}) + \{\mathrm{CO}\} \tag{6.2}$$

$$[\mathrm{Fe}] + C_{gr\ \text{from electrode}} = [\mathrm{Fe_3C}] \tag{6.3}$$

When the arc is not burning, the electrodes wear solely due to the oxidizing effect of the furnace atmosphere, or if the electrodes are placed outside the furnace (e.g., during charge loading), the ambient atmosphere. Due to the small area of the electrode face in relation to its total surface as well as fairly short periods of the electrodes remaining in such a state, the percentage share of this consumption in the overall balance is small.

In general, it can be found that the wear of the electrode tip face largely depends on the diameter of its tip. This, in turn, primarily results from the wear of the side surface of the electrode. Based on many research studies and industrial observations, it can be assumed that the intensity of the linear electrode consumption increases as the arc current increases.

6.2.2 Wear of Side Surface of the Electrode

The side surface of the electrode wears in the vertical direction, causing it to converge in the part below the furnace roof (the part located directly in the furnace vessel). The oxidation is the most important factor contributing to the side surface wear. The destruction of this layer is also caused to some extent by the erosive and corrosive effects of the condensed vapors of the components of the metal bath and the slag. The resulting liquid condensate droplets, as they flow down the electrode, "take" some amounts of graphite with them. The oxidation of graphite electrodes mainly depends on the temperature as well as on the area of the electrode surface, the content of oxidizing components in the gas atmosphere in the furnace, the degree of their turbulence, and the quality of the electrodes. The process also depends on the residence time of the electrode inside the arc furnace vessel [6]. The most important factors influencing the rate of oxidation of the side surfaces of the electrodes are as follows:

- chemical composition of the gas atmosphere inside the vessel, especially its oxygen content,
- electrode surface temperature,
- flow rate and degree of turbulence of gases moving around the electrodes,
- electrode quality.

This is confirmed by the results of the tests obtained in the wind tunnel as shown in Figure 6.4.

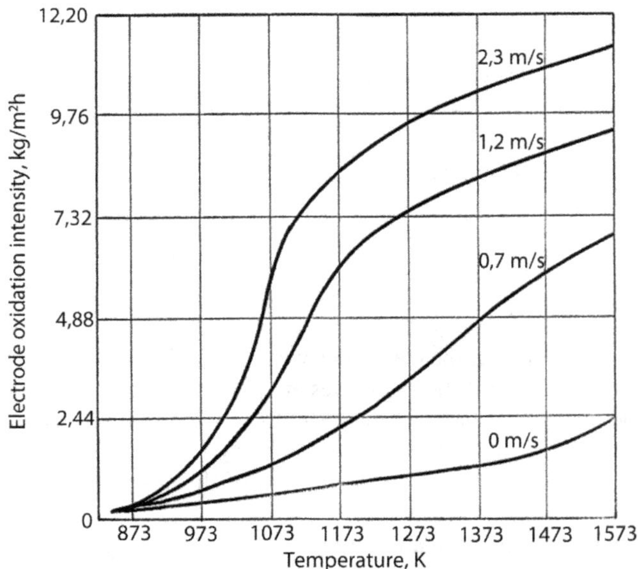

FIGURE 6.4 Oxidation intensity of the side surface of a graphite electrode versus the temperature and flow rate of gases with a composition typical for an arc furnace atmosphere. (Author's work.)

The oxygen content in the furnace atmosphere is similar to that in the air. The oxidation rate of the electrode side surfaces also depends to some extent on the crystalline structure of the graphite. The wear of the electrode side surface is directly proportional to the furnace atmosphere impact time. Long times may lead to pronounced thinning of the lower parts of the electrodes, which is often misinterpreted as a result of their inadequate resistance to oxidation. Short residence times of the electrodes in the furnaces, due to the intense linear (tip) wear, result in a smaller taper of the lower part of the electrode. The types of chemical reactions according to which the electrode side oxidation process takes place depend on temperature. The following reactions are assumed [8]:

$$xC + yO = C_xO_y \quad \text{for} \quad T < 723\ \text{K} \tag{6.4}$$

$$C + O_2 = CO_2 \quad \text{for} \quad T < 1{,}073\ \text{K} \tag{6.5}$$

$$C + \frac{1}{2}O_2 = CO \quad \text{for} \quad T > 1{,}073\,\text{K} \tag{6.6}$$

$$C + CO_2 = 2CO \quad \text{for} \quad T > 1{,}773\ \text{K} \tag{6.7}$$

In order to reduce this type of electrode consumption, protective coatings that are immune to the effects of the oxidizing atmosphere are used. Many solutions have been developed in this respect, and a few of them are used in practice. Electrodes with coatings like these are characterized by a significantly lower consumption in the production practice [9].

During the oxidation of the electrodes, the heat equivalent to 10–20 kWh per metric ton of steel is released. This heat also heats the electrodes and the space inside the furnace vessel. Water is sprayed in order to reduce the oxidation of the side surface of the electrodes. The spraying is performed by placing rings made of steel pipes around the electrodes directly above the roof. Small holes are made in the rings, from which cooling water flows under pressure, creating jets directed at the side surface of the electrodes. The water flowing down their surface evaporates, and the energy needed for evaporation cools the electrodes. The lowered temperature of the electrodes significantly reduces the wear of their side surface. The relative share of the linear wear of the electrode tip and the oxidation of their side surface in the total wear of the electrodes (not taking into account fractures, cracks, and butt losses) is characterized by the graphs shown in Figure 6.5.

The graph shows that, for a small ratio of the electrode tip face diameter to the nominal diameter, the side surface wear is the factor limiting the process-related wear, while for large quotients of these diameters, the limiting factor is the wear of the electrode face. In industrial conditions, the quotient *d*/*D* is usually within the range of 0.6–0.8. Therefore, as the graph shows, both factors are more or less of equal importance.

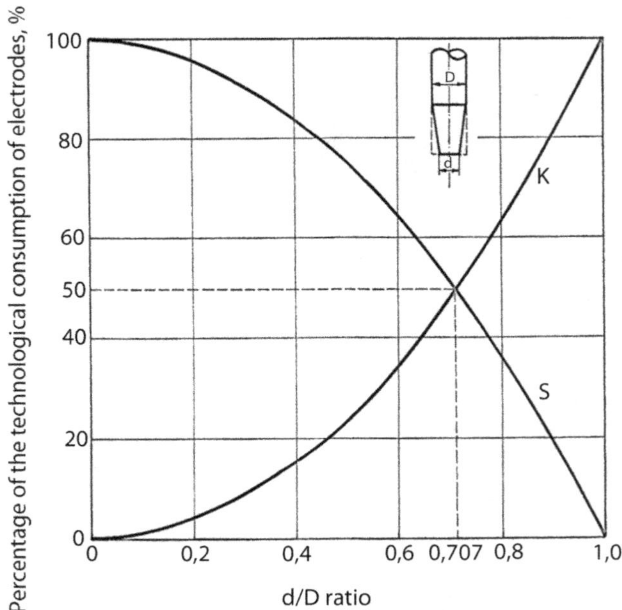

FIGURE 6.5 Percentage share of the process-related wear of the tip and side surface versus the ratio of the electrode tip diameter to the nominal diameter (total wear is $C=K+S$, where the tip consumption is $K = c_1\left(d/D\right)^2$ and the side surface wear is $S = c_2\left[1-\left(d/D\right)\right]^2$. (Author's work.)

6.2.3 Other Reasons for Electrode Consumption

This category of causes of consumption of graphite electrodes working in arc furnaces includes all of the phenomena causing their sudden shortening. A distinction is made here between the loosening of the electrodes at joints, cracking, and breaking. The reasons for loosening the electrodes at joints can be classified as follows: mechanical vibrations of the columns caused by a rotating electric arc, continuous operation of the electrode feed automatic control system, electrodynamic phenomena, caving of an improperly arranged charge, incorrectly selected dimensions of connecting pins and sockets, and incorrect assembly of the column. In order to protect the electrode column from falling apart during the operation, various solutions are used, generally referred to as "locking."

Cracks can be another cause of electrode sudden shortening. In the operating conditions of the arc furnace, splitting or so-called "V" cracks may appear. Splitting is caused by excessive thermal shock when withdrawing the electrodes from the furnace vessel. The cause of "V" cracks is the formation of a significant temperature gradient on the electrode surface resulting from the prolonged arcing in one place, which can occur, for example, when melting thick scrap.

Breakage of the electrode columns is primarily observed during scrap meltdown. The primary cause of breakage is caving scrap, which can form a kind of support for the electrode column. It can also happen – when the charge is improperly loaded – that the electrode encounters a nonconductive material and rests against it, and the automatic control system pushes it down with a great force, thereby causing a breakage. In addition, significant electrodynamic forces impact the electrode column during the meltdown, resulting from the maximum arc powers and high currents applied during this period. The value of these forces depends on the square of current intensity.

The percentage share of individual causes of wear of graphite electrodes (calculated in industrial conditions as the wear of electrode in kg per ton of steel) operating in typical conditions of a steelmaking arc furnace is: approximately 95% accounts for process-related wear and approximately 5% for accidental wear.

6.3 INFLUENCE OF THE OPERATING PARAMETERS AND STEELMAKING CONDITIONS ON ELECTRODE CONSUMPTION

The electrode consumption mechanisms discussed above show that the basic operating conditions of the EAF influencing the life of graphite electrodes are electrical energy parameters (the so-called electrical load) and the tap-to-tap time. The electrical parameters primarily affect the wear of the electrode tip, and the tap-to-tap time affects the oxidation of the side surface. The influence of other factors, such as the composition of the gas atmosphere inside the furnace vessel, steelmaking practice applied, and number of breakages or cracks, is of minor importance.

The findings of industrial research show that the relationship between the electrode consumption and electrical parameters as well as the melting time can be described by the following statistical relationship [6]:

$$q = k \cdot I \cdot t \tag{6.8}$$

where:

- q – electrode consumption [kg/t of steel],
- k – numerical factor, the value of which depends on furnace operating conditions and which is $(16–22)\cdot 10^{-6}$ for electrodes of 400 mm diameter; $(5–11)\cdot 10^{-6}$ for 500 mm electrodes; $(4–5)\cdot 10^{-6}$ for 610 mm electrodes,
- I – current during the meltdown [kA],
- t – tap-to-tap time [h].

Some data show that, in a relationship (6.8), the exponent of the current may appear with the power of two or even three. This is especially true of large super-ultra-high power (SUHP) furnaces, with installed high capacity transformers, where high unit electric power is used. Then, the tap-to-tap time is short, and the consumption of electrodes depends on the current rather than on the melting time. Note that the electrode consumption is directly influenced by the current, but not by the voltage. Therefore, in order to minimize the consumption of electrodes, it is advantageous to use such operating states of the power supply system that allow long arcs to burn. A specific electrical energy can be converted into heat energy either in a long arc, which is created with a lower current and a higher voltage, or in a short arc, which is generated with a higher current and a lower voltage.

While the process-related consumption of electrodes is a natural phenomenon related to the melting process, the accidental consumption usually results from the improper preparation of joints in the electrode columns as well as the operation of the electrode control system with low precision or the improper preparation of scrap loaded into the furnace vessel. These are random factors, dependent on the work of the furnace operators. They include breakages, cracks, and so-called butt losses.

The electrodes usually break near the top joint (closest to the column holder). The breakage occurs at the level of the joint socket below or above the nipple or in the half of the joint length. The primary cause of electrode breakage is the caving scrap, which forms a kind of support or impacts the horizontal direction causing the electrode to bend. Of particular note is that the electrodes are also affected by electrodynamic forces directed outward from the axis of the furnace. These forces increase as the square of the current intensity increases, and they decrease as the distance between the electrodes increases. In large furnaces, where the current intensity is over 50 kA, these forces reach a value of tens to hundreds of newtons. The occurrence of these forces requires the use of an appropriate method for connecting the electrodes as well as the proper design and operation of the system controlling the movement of the electrode arms.

The term "butt losses" is understood as a fall-off of the last piece of the electrode, located immediately in front of a joint. The causes of butt losses include loose joints caused by improper workmanship or excessive column vibrations. A butt, sometimes weighing more than several dozen kilograms, drops into the metal bath after falling off, which causes a significant disruption to the melting practice due to a sudden increase in the amount of carbon in the liquid metal.

Excessive thermal shocks occurring when the electrodes are withdrawn from the furnace vessel can create shear stresses on the surface, which under extreme conditions can even cause the electrode to split along its entire length. When heavy scrap is melted, a stationary arc is formed at the outer edge of the tip, thereby causing large temperature gradients. This is accompanied by high internal stress making "V" cracks propagate to the nearest joint. Depending on the depth of such a crack, the butt may also split and fall off.

Whatever the cause of a sudden shortening of an electrode, it practically always causes a disturbance in the steelmaking process. Therefore, the quality of the electrodes, especially the bending strength, and the responsible work of the furnace operator, including the appropriate preparation of the scrap in the charge basket as well as the correct assembly of the electrode sections into a column, determine the correct steel production process.

The consumption of electrodes is one of the most important indicators of an arc furnace operation. Due to a relatively high price of electrodes, the consumption rate (expressed in kilograms of electrodes used per 1 ton of steel produced) is a decisive element in the costs of steel production. The electrode consumption rate in modern high-power furnaces with a short meltdown time is under 1 kg/t; in older design furnaces with a low installed power of the furnace transformer, it reaches 5 kg/t, and in some special cases, it may exceed this value [10].

REFERENCES

1. Armstrong C.: Electrode Consumption in a Modern High-Power Electric Steel Shop, *Iron and Steel Technology*, Vol. 1, 2013, pp. 57–59.
2. Taylor C. R.: *Electric Furnace Steelmaking*, The Iron and Steel Society, Warrendale, PA, 1985.
3. Kohle S.: Improvements in EAF operating practices over the last decade, *Proceedings of 57th Electric Furnace Conference*, Pittsburgh, PA, 14–16 November 1999.
4. Cozzi G., Lombardi E., Ferri M.: Reduction of Energy and Graphite Electrode Consumption at ORI Martin, *Metallurgical Plant and Technology International*, Vol. 26, 2003, pp. 40–42.
5. Rafiei R., Kermanpur A., Ashrafizadeh F.: Numerical Thermal Simulation of Graphite Electrode in EAF During Normal Operation, *Ironmaking and Steelmaking*, Vol. 35, 2008, pp. 465–472.
6. Karbowniczek M., Krucinski M.: Wplyw stanow pracy pieca łukowego na zużycie elektrod grafitowych, *Hutnik*, Vol. 9, 1985, pp. 273–276 (in Polish).
7. Liu G., Zhu R., Lu D., Jiang G.: Study of Mechanism and Technique of Lowing EAF's Graphite Electrode Consumption, *Metallurgical Equipment*, Vol. 3, 2008, pp. 56–59.
8. Krucinski M., Karbowniczek M.: Zużycie elektrod grafitowych jako funkcja stanow pracy pieca lukowego, Zeszyty Naukowe AGH, *Chemia*, Vol. 1220, 1989, z.10, s.61–s.70 (in Polish).
9. Friedrich C.: Graphite Electrodes Climb the Diameter Ladder: Larger Diameter Electrodes Can Provide Greater Productivity to an EAF by Reducing the Number of Electrode Changes Required, *Steel Times International*, Vol. 29, 2005, p. 25.
10. Martell F., Mendoza R., Melendez M., Llamas A., Micheloud O.: Increasing Energy Efficiency of the Electric Arc Furnace at Tenaris Tamsa, *Iron & Steel Technology*, Vol. 10, 2013, pp. 81–89.

7 Environmental Protection Systems

The implementation of technological operations in a modern steelmaking arc furnace, combined with methods that intensify the processes, results in certain harmful effects on the environment. The most important impacts include the generation of significant amounts of process gases and dust (so-called off-gases or waste gases), noise, and high temperatures in the vicinity of the furnace. In order to eliminate all these environmental burdens, or at least significantly reduce them, various technical and structural solutions are used in steel plants. The aim of such solutions is to ensure safe and environmentally friendly working conditions for the furnace operator, as well as to ensure that the emission of environmentally harmful substances is within the limits allowed by the regulations. For these reasons, the design of furnace and auxiliary devices must provide effective systems for gas and dust extraction, noise protection, and protection against high temperatures.

7.1 PROTECTION AGAINST GASES AND DUSTS

Undoubtedly, from the point of view of environmental protection, the most important is the system of extraction of gases and dusts generated in the metallurgical process from the furnace [1–4]. Its task is to maximally capture gases from the working space of vessel and post-combustion of CO in them, in addition to cooling down and filtering out the dust. During the smelting of steel in the arc furnace, as well as during the charging and tapping operations, considerable amounts of gases contaminated with dusts are released, which could pose a threat to the natural environment in the vicinity of the steelworks and directly to the operating personnel. This is of particular importance in the case of modern furnaces, where technologies intensifying the smelting process employing gaseous oxygen, gas-oxygen burners, or slag foaming technology are used. The choice of the construction of the off-gas system and the parameters of its operation determine the effectiveness, i.e. the content of undesirable components emitted from the technological process outside the furnace – into the atmosphere. High efficiency and reliability of gases capture can be achieved with an advanced system of monitoring and control of parameters in the furnace, such as pressure. Proper control of the quantity of captured gases is of key importance for the reduction of dust emissions, furnace efficiency, electricity consumption, electrode consumption, and other work indicators, which translates directly into the economy of steel production.

In the classic steel smelting process, the amount of gases emitted is 20–50 Nm^3/h, per 1 ton of furnace capacity. The use of gaseous oxygen to refine the metal bath increases the volume of gases released to 200–400 $Nm^3/h/ton$ of furnace capacity, with a significant increase in the amount of dust. In currently operating furnaces,

DOI: 10.1201/9781003130949-7

equipped with gas-oxygen burners, in which the technology of intensifying the melting of charge using gaseous oxygen and slag foaming, the volume of gases released is in the range of 500–1,000 Nm^3/h, per ton of furnace capacity. The gases released from the furnace vessel, depending on the quality and type of charge materials used, the unit power of the furnace transformer, the technology used, and the smelting time, contain dust in the amount of 50–100 g/m^3, which many times exceeds the permissible limits (usually the allowed dust content does not exceed 40 mg/m^3). The dust from electric arc furnaces (EAFs) is extremely dispersive as approximately 90% of the particles are smaller than 0.5 μm.

The main components of the waste gases are carbon monoxide, carbon dioxide, nitrogen, and oxygen. Carbon monoxide is produced as a result of chemical reactions of carbon oxidation, and carbon dioxide as a result of the reduction of iron oxides in the slag, post combustion of CO and the decomposition of carbonates contained in the charge. The total amount of $CO + CO_2$ in the waste gases usually does not exceed 30% and significantly depends on the smelting stage, the type of technology used (including the amount of oxygen applied), the scope of use of gas-oxygen burners, etc. The rule is that if the share of CO in the gases increases, then the share of CO_2 decreases. Both gases appear in maximum amounts during the melting of the charge, while in the remaining technological periods, their share is smaller and amounts up to several percentage. At that time, mainly CO_2 emitted during slag foaming occurs in the waste gases. The oxygen content in the waste gases is also related to the technological stages of smelting and varies from practically zero to about 20%. During the charge loading and the bath reheating as well as refining stage, the share of oxygen in the waste gases reaches 20%, but during the charge melting its share is the smallest. The remaining part of the gases generated inside the furnace vessel is nitrogen, coming from the air sucked in from the environment as a result of the negative pressure generated by the exhaust system. Examples of changes in the chemical composition of gases inside the furnace vessel are shown in Figure 7.1.

When wet scrap or oiled shavings (containing hydrocarbons) are loaded into the furnace, insignificant amounts of hydrogen and organic compounds, such as benzenes, phenols, aromatic hydrocarbons, chlorobenzenes, polychlorinated biphenyls and polychlorinated dibenzodioxins or dibenzofurans, may appear in the waste gases. The amount of organic compounds is at the level of 0.1–0.5 ng I-TEQ/Nm^3 [5,6]. Although in small amounts, sulfur oxides as well as fluorides may also occur along with nitrogen oxides and ozone, which are formed in the area of the burning electric arc, where conditions for gas ionization occur.

The main components of the dust are iron oxides, formed when oxidation of the metal bath with gaseous oxygen occurs. Another important component is zinc oxide, formed when galvanized steel scrap (mainly sheet) is used as a charge. At temperatures of the steelmaking process (in the range of 1,873 K (1,600°C)), zinc evaporates and is sucked in as vapor with the waste gases; it then oxidizes and in the form of solid ZnO particles becomes a component of dusts. The share of zinc oxides in the dust, depending on the amount of galvanized steel scrap used as a charge, may vary widely, from 5% to 50%. The other dust components are MnO, CaO, and SiO_2, in amounts up to a few percentage as well as in amounts not exceeding 1%: Na_2O, K_2O, PbO. In case of melting steel scrap containing chromium, in the dusts, some

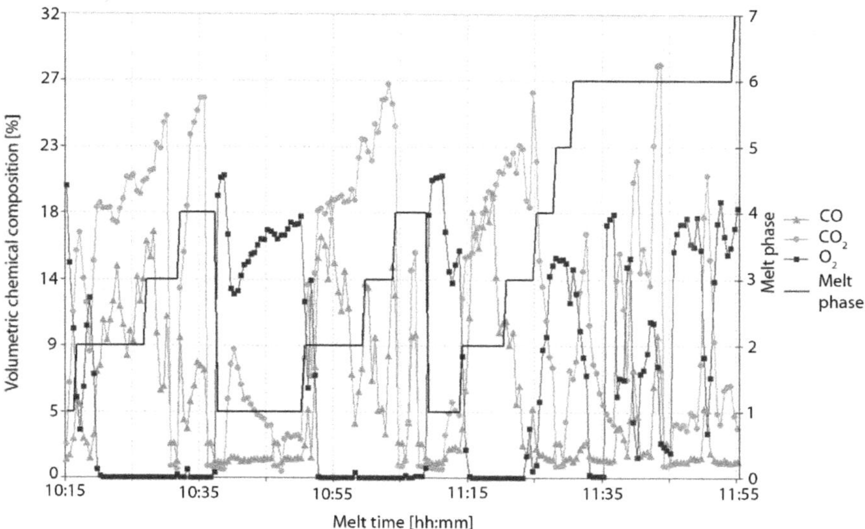

FIGURE 7.1 Examples of changes in the chemical composition of gases inside the furnace vessel, sucked into the exhaust duct; smelting phase: 1 – charge loading, 2 – initial melting phase, 3 – melting + oxygen lance, 4 – melting + oxygen lance and foaming slag, 5 – sampling and temperature measurement, 6 – tapping, 7 – end of heat. (Author's work.)

quantities of Cr_2O_3 may appear. Depending on the method of loading the charge, and in particular the way of feeding the slag-forming materials, the quality of the materials loaded, the amount of oxygen used, and the range of slag foaming technology, several to about 20 kg of dusts are generated per 1 ton of steel produced [7].

Waste gases can be considered in two emission categories: primary and secondary. The primary gases are those that are emitted during the melting of the charge and metal bath refining; they account for about 90% of the total gas emissions from an EAF. Secondary gases are generated mainly during charge loading, sampling, temperature measurement, and tapping. Secondary waste gases, which account for about 10% of the gases emitted from the furnace, are created in short periods of time, e.g. when the roof is raised and moved from above the furnace. Similarly, dusts contained in the gases, both in terms of quantity and chemical composition, are correlated with emission categories. The main amount of dust in the basic emission is 95%. In the secondary emission, however, no more than 5% of the dusts are contained [8,9].

7.2 GAS CAPTURE SYSTEM

Initially, the system to reduce the emission of gases and dusts into the atmosphere included only the hood above the furnace, connected by a pipeline with fans, and then through a filter station with a chimney. Such a system did not guarantee sufficiently good gas and dust capture efficiency. The systems in which gases are captured through the so-called fourth hole in the roof are more effective. Such a

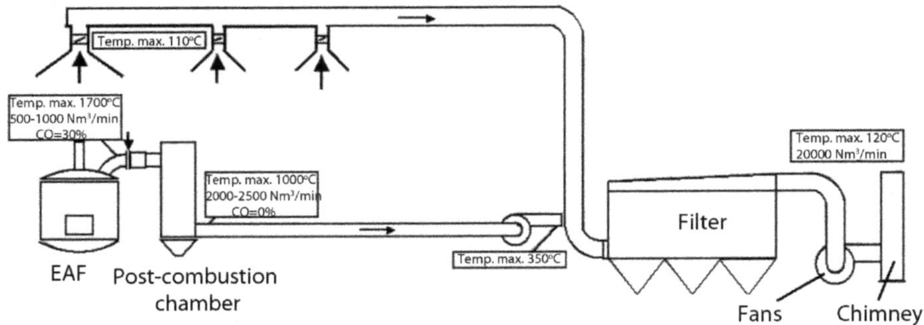

FIGURE 7.2 General scheme of off-gases capture system from the 160 tons electric arc furnace. (Author's work.)

solution is still a commonly used way of "capturing" the waste gases, regardless of other additional elements of the whole system. The general scheme of the system of dusty gases capture from the EAF is presented in Figure 7.2. The system includes the capture of gases in two ways. The basic way is to capture the gases through the fourth hole in the furnace roof. Gases and dusts not captured in this way, escaping through leaks at the contact between the roof and vessel walls and through leaks in the roof holes for electrodes (larger diameter of the holes than the diameter of the electrodes to allow the electrodes to move), are taken over by hoods placed over the furnace and over other devices, e.g. secondary metallurgy devices.

In Figure 7.2, in the lower left corner, there is an EAF generating 500–1,000 Nm^3 of gas per minute. The maximum temperature of the gases is 1,973 K (1,700°C), and they usually contain up to 30% of CO. The majority (it can be assumed that a minimum of 90%) of the gases and dust generated in the process are sucked through the fourth hole in the roof, on which the so-called over-roof elbow is installed. Another element of the system is the so-called post-combustion chamber. Between the elbow outlet and inlet to the chamber there is an air gap, usually of adjustable width. Cold air is sucked in through this gap, which allows the supply of oxygen to post-combustion CO. The width of the gap, and thus the regulation of the amount of oxygen supplied, is controlled depending on the amount of CO in the waste gases and their temperature. The air intake also contributes to the initial cooling of the gases. The reaction of CO to CO_2 combustion takes place in the post-combustion chamber. For this purpose, nozzles for blowing oxygen or air are sometimes installed in the chamber. The combustion reaction is exothermic, which increases the gas temperature. The cooling aspect of the intake air is usually greater than heating aspect of CO combustion, hence the average temperature of the gases in the post-combustion chamber decreases. The post-combustion chamber is made as a steel structure, often in the shape of a cylinder, from water-cooled elements. Sometimes this part of the extraction system is made in the form of a drum, with the possibility of heat recovery from the waste gases. Then the water heated in the drum or the obtained steam become the heat carrier transferred for heating purposes either directly in the steelworks or commercially outside. The post-combustion chamber is equipped in its lower part with a conical tank in which thicker dust fractions accumulate due to gas swirls; it can be assumed that it is the first stage of cleaning gases from dust. As shown in the figure, the amount of waste gases, after leaving the post-combustion chamber, increases and is in the range of 2,000–2,500 Nm^3/min.; with a maximum temperature of 1,273 K (1,000°C) and practically with no CO content.

The pipeline transporting gases from the post-combustion chamber toward the filter station is another element of the off-gas system. A pipeline with round or rectangular cross section is made from steel sheets. The length of the pipeline is 100–300 m, and such a length aims at gas cooling. Usually no special cooling of the pipeline is used although sometimes its initial part is made as water cooled. The temperature of gases at the end of the pipeline should not exceed 623 K (350°C). The suction of gases and their subsequent flow through the off-gas capture system are forced by a set of fans located at the end of the pipeline. For precise control of the quantity of gases sucked in from the furnace, the fan system is equipped with dampers, which are characterized by a variable geometry of the opening degree. In this way, depending on the stage of the smelting process, it is possible to control the negative pressure in this section, which determines the gas flow rate. The waste gases behind the fans are often still too hot to be passed on to the final dust collector. For this purpose, a technique of mixing them with cooler waste gases sucked in through hoods is frequently used.

This part of the off-gas capture system through the hoods is shown in the upper part of the figure. On the left side, there is a hood directly above the furnace, then there are examples of two further hoods that can be installed in the area of steel tapping from the furnace – over the ladle, over secondary metallurgy devices, etc. The hoods allow the suction of gases from the above-mentioned stations. Also, for example, when the charge is loaded into the furnace, the roof is lifted and deflected together with the elbow, and the gases and dusts escaped during this time are sucked through the hood. The temperature of the gases sucked into the hoods, as a rule, does not exceed 373 K (100°C). From the hoods they are transported through a common duct toward the off-gases capture devices located outside the steelworks. Fabric bag filters are most often used in off-gases capture facilities for arc furnaces; sometimes electrostatic or wet filters are applied. Fabric filters are built in the form of a rectangular steel tank with a set of filters inside, made from temperature and chemically resistant fabrics. The waste gases after cleaning from the dust are emitted to the atmosphere through a chimney. The capacity of the fans after the filter station should be up to 20,000 Nm^3/min.

The general scheme of connections of the off-gases capture system in the electric steelworks is presented in Figure 7.3. The scheme shows the system including the off-gases capture of the arc furnace, the station for unloading and preparation of charge materials, and the ladle furnace, with the possibility to process filtered dusts. The general scheme of connections between the arc furnace and the off-gases capture system is shown in Figure 7.4.

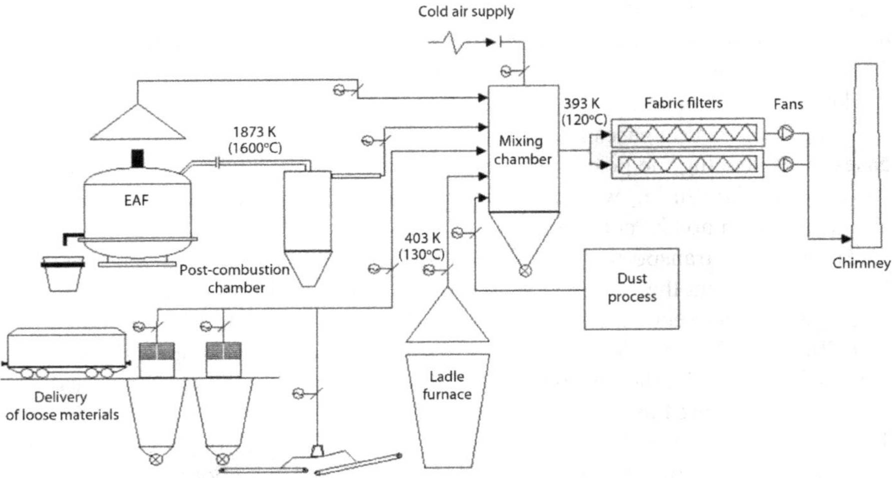

FIGURE 7.3 General overview of the connection scheme of the off-gases capture system from the arc furnace vessel. (Author's work.)

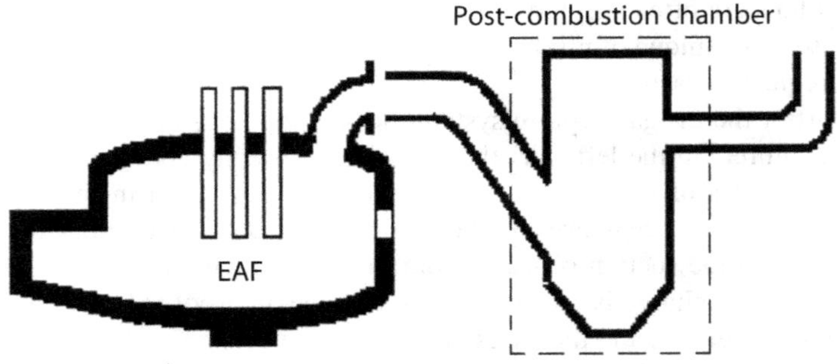

FIGURE 7.4 General overview of connections between the arc furnace and the off-gases capture system. (Author's work.)

7.3 MODELING OF THE SYSTEM OF OFF-GAS CAPTURE FROM THE ELECTRIC ARC FURNACE

Modeling of the off-gas capture system from an EAF is one of the aspects of the analysis of its operation, and it enables the optimal design of the structure and selection of the most advantageous technical parameters of the system and technological parameters of the smelting. One of the modeling methods is the application of computational fluid dynamics (CFD). Based on the design parameters of the gas capture system, a computational model of the furnace channel can be defined. An exemplary structural model of the system, for a furnace with a capacity of 160 t, is presented in Figures 7.5 and 7.6. For such a system, computational parameters have been defined, including gas flow characteristics, type of computational grid,

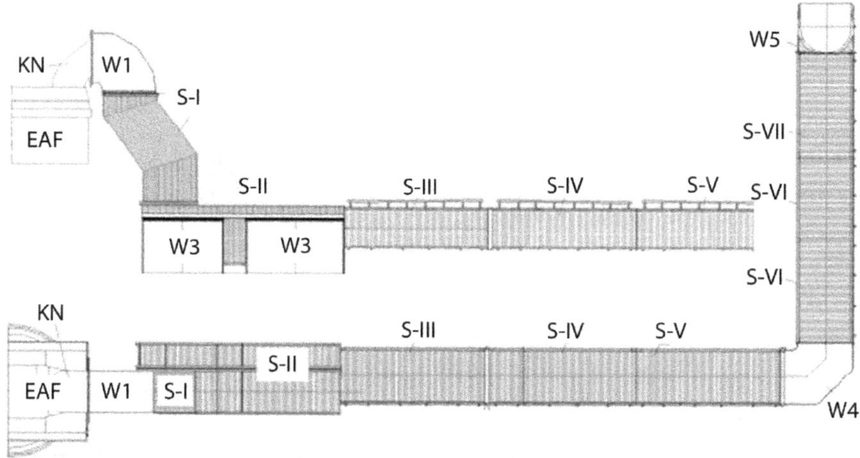

FIGURE 7.5 General scheme of a structural model of an exemplary off-gases capture system from the arc furnace vessel [12]. (Source: Karbowniczek and Kawalkowski 2007.)

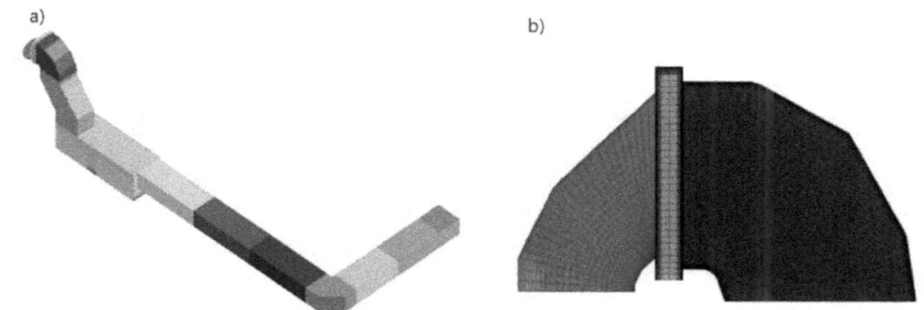

FIGURE 7.6 Spatial structural model of an exemplary off-gases capture system of an arc furnace (a) and an air gap between the over-furnace elbow and the sliding elbow (b) [12]. (Source: Karbowniczek and Kawalkowski 2007.)

boundary conditions, and input assumptions [10–12]. Commercial software CFD – FLUENT code was used. The flow-thermal character of the simulation in the model was adopted. The temperature of gases at the inlet to the system of 1,873 K (1,600°C) was assumed along with their chemical composition: CO –19.3%, CO_2 – 19.0%, O_2 – 0%. During the simulations, the temperature distribution, flow velocity, and carbon monoxide content in the cross section and longitudinal section of the modeled channel were calculated. The results are presented graphically in Figures 7.7–7.10. The subsequent figures show the calculated spatial gas temperature distribution, flow velocity distribution, and the percentage distribution of CO_2 and CO inside the initial part of the capture system. Figure 7.7 also shows the actual temperatures measured at three points. The results of simulation calculations may also be helpful in controlling the technical parameters of the "on-line" capture system. For example, based on the measured temperature level and the content of carbon monoxide and dioxide

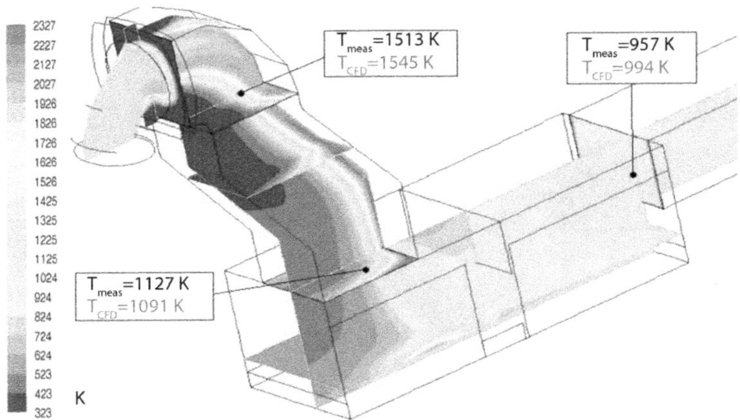

FIGURE 7.7 Calculated spatial distribution of gas temperature inside the initial part of the capture system together with the actual measurements given in three points [12]. (Source: Karbowniczek and Kawalkowski 2007.)

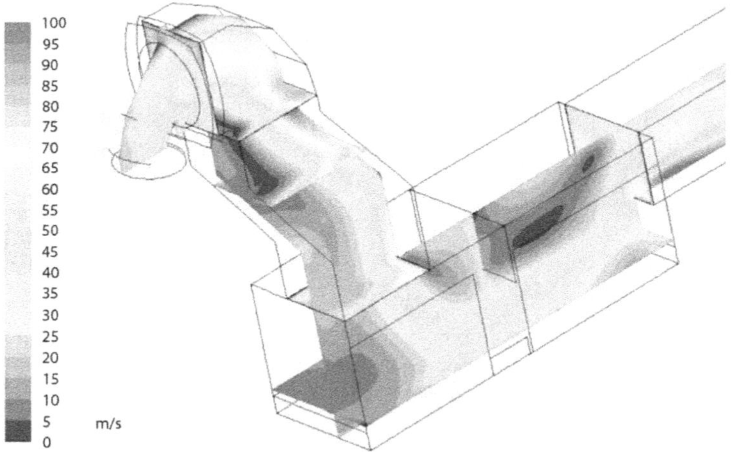

FIGURE 7.8 Calculated spatial velocity distribution of gases inside the initial part of the capture system [12]. (Source: Karbowniczek and Kawalkowski 2007.)

at the outlet of the elbow, it is possible to control the operation of burners, oxygen, and carbon lances, as well as the post-combustion lances in the post-combustion chamber. The negative pressure in the capture system can also be controlled, which determines the amount of sucked gases, adapted to the current needs resulting from the metallurgical process. The executive element in this control system is to change the position of the dampers in the capture system.

The results obtained illustrate the behavior of the gases during flow in the channel of the capture system. In the case under consideration, the temperature distribution in the initial part of the system shows significant differences, i.e. in the upper part and

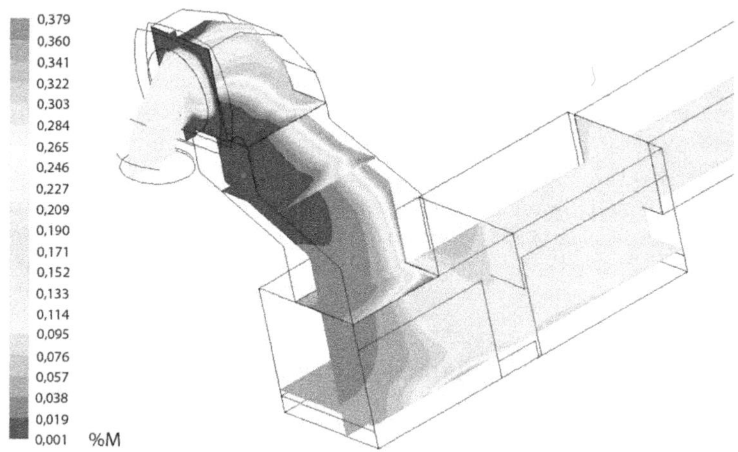

FIGURE 7.9 Calculated spatial distribution of the percentage of CO_2 inside the initial part of the capture system [12]. (Source: Karbowniczek and Kawalkowski 2007.)

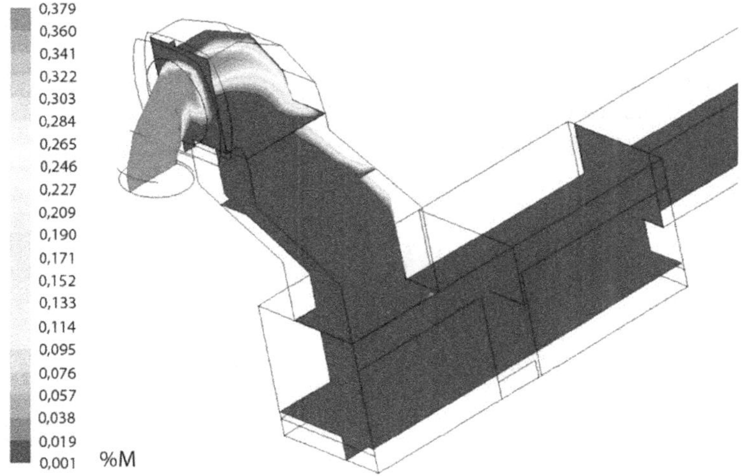

FIGURE 7.10 Calculated spatial distribution of the percentage of CO inside the initial part of the capture system [12]. (Source: Karbowniczek and Kawalkowski 2007.)

in the right side of the elbow, the temperature reaches over 1,873 K (1,600°C), while in the lower left side, as a result of sucking cold air through the gap, the temperature is about 573 K (300°C). In the horizontal part of the system, the temperature in the whole section of the channel equalizes and reaches the values of 973 K (700°C). Similar in terms of uniformity, spatial distribution was obtained for the gas flow velocity as well as the percentage of CO_2 in the channel. On the other hand, a completely different spatial distribution was obtained for the percentage share of CO in gases – already in the elbow there is practically a complete decrease of its share to zero, as a result of afterburning to CO_2.

Analogous results of gas flow parameters in the off-gases capture system of the arc furnace are obtained for other furnace sizes and other detailed solutions of their construction. The simulations and analyzes of numerical calculations, in connection with industrial verifications, enable the optimization of the working conditions of the systems, especially in relation to the minimization of the harmful environmental impact of the production process in the arc furnace [13]. At the same time, modeling of these systems facilitates the design of new installations in order to obtain the most favorable operating conditions.

7.4 COMPREHENSIVE FURNACE HOUSING SYSTEM

Modern furnaces use comprehensive systems to isolate the furnace from the surrounding environment. Such a system includes the enclosure of the entire furnace in a special housing that allows complete isolation of the working furnace from the surrounding environment, in terms of dust, noise, and thermal conditions. Two types of such systems are used: a so-called dog-house (covering the furnace itself) or a so-called elephant-house (covering the entire furnace including auxiliary equipment). Regardless of the type, the housing is made from steel sheets lined with sound-absorbing material inside. The housing has a sliding element – the so-called gate through which the scrap is loaded and the access for the furnace operator is provided. The scheme of the "elephant-house" is shown in Figure 7.11.

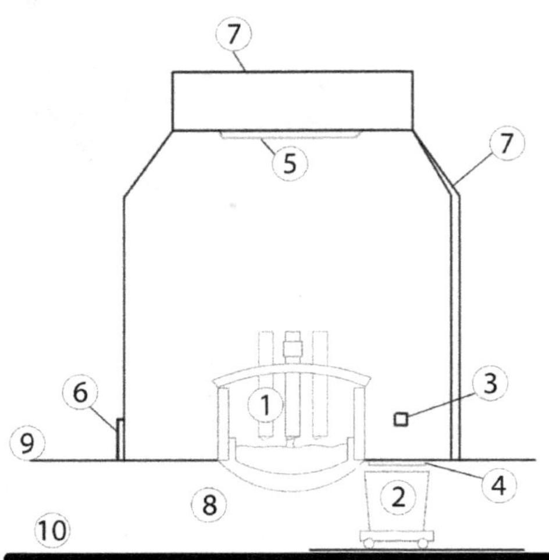

FIGURE 7.11 General illustration of the scheme of the "elephant-house" furnace housing: 1 – arc furnace, 2 – ladle, 3 – charging chute for metallic additives, 4 – gas capture duct, 5 – main gas capture duct, 6 – side gate, 7 – main loading gates, 8 – slag storage yard, 9 – working level, 10 – bay level. (Author's work.)

Such a housing enables complete capture of gases and shields the surroundings and all workstations, including the furnace platform, against noise. The shielding wall is constructed of two sheet metal shells with a sound-absorbing material between them. The total wall thickness is 400–500 mm. The noise level is reduced to the level of 25–40 dB. The furnace vessel gases are captured through a fourth hole in the roof. The remaining gases and dust generated during post-tapping repair, charge loading, tapping, and other operations are captured by the hood located under the roof of the housing. Due to the fact that the volume to be dedusted is much smaller compared to the open furnace system, the volume of the gases to be dedusted is relatively small, i.e. approximately 40% of the gas volume in the classic solution.

The gate to the furnace housing has a sliding structure. When loading the charge, the gate is moved away, exposing the vessel and enabling loading. At the same time, it is possible to perform other auxiliary operations, such as electrode column replacement, post-tapping repair, etc. After all the preparatory operations are completed, the gate is closed before the start of the melting scrap. The entire melting period is carried out with the gate closed, which enables the smelting process to be completely isolated from the surroundings. Under the working platform there is a similar, smaller chamber for a vehicle with a ladle for steel and a vehicle with a ladle for slag, which allows for full capture of gases during tapping and slagging. It is also possible to use the chamber for argon refining of steel in a ladle placed on the vehicle. The resulting gases are captured by the general dedusting system. Gases are extracted near the top of one of the walls of the housing, and fresh air is introduced from below through openings located in the working platform.

The efficiency of gas removal and dedusting in this system is higher due to the smaller volume of space in which all sources of contamination are enclosed. The disadvantage of this technique is the more complicated and time-consuming operation of the furnace and the need for higher investment outlays (e.g. for additional gate closing and opening mechanisms, and charging and tapping procedures) [14].

Noise protection is the second important aspect of using a furnace housing. The main source of noise emission in a steelworks is the EAF, and in particular the sounds that occur during a burning electric arc, especially in the initial phase of melting cold scrap. The arc then burns very unstably with frequent interruptions and reignition, which, in turn, causes loud discharges. Apart from the furnace, other auxiliary devices, such as fans, burners, carriages, cranes, and scrap falling during loading, can be a source of noise. The noise level in the steelworks at this time reaches the value of 120 dB, and sometimes even more. This significantly exceeds the applicable standards, which in most countries allow industrial noise levels to a maximum of 90 dB, once every 8 h, and in no case exceeding 115 dB. Such noise level, like fumes and dust, is a health hazard for the steelworks staff and its surroundings.

The use of an acoustic housing on the EAF can reduce the average sound intensity by 20–50 dB. The housings can also be used in secondary metallurgy processes, but this requires a special steelworks wall design to eliminate echo. The furnace housing devices, apart from the aspect of minimizing dust emission and noise reduction, also enable the reduction of thermal radiation energy to the environment. All this contributes to the improvement of the working conditions of the furnace operators and the minimization of environmental hazards [14].

7.5 USE OF WASTE GAS THERMAL ENERGY TO HEAT UP SCRAP

For over 30 years development works have been carried out to increase energy efficiency of the process of steel smelting in an arc furnace [15,16]. One of such activities is the concept of using the thermal energy of waste gases to heat up scrap before it is loaded into the vessel. The aim is to increase energy efficiency, which will enable reduced total electricity consumption. Research works have been and are being carried out in this area, pilot installations on an industrial scale have been designed and launched, but so far, no satisfactory concept has been found that could be universally implemented. The work focuses on the use of a chamber located directly above the furnace or next to the furnace, in which cold scrap is placed before loading and into which hot gases are introduced directly from the vessel.

One of the simpler solutions in this area is heating up the scrap in the charging basket. The general concept of such a system is presented in Figure 7.12. In the waste gases captured system, a chamber made from steel is then installed, with an insulating lining inside, to which hot exhaust gases from the furnace vessel are fed. A scrap charging basket is placed in the chamber, before loading into the furnace vessel. It can be assumed that the temperature of gases fed into the chamber where the charging basket will be placed will be about 1,073 K (800°C) and those leaving about 473 K (200°C). The degree of heating up depends on the size of scrap pieces and the heating time. The temperature of the scrap after heating should be between 573 K (300°C) and 723 K (450°C). Such a temperature of scrap before the start of melting will allow saving electricity by about 40–60 kWh/ton of steel, reducing the consumption of electrodes by about 0.3 kg/t, and shortening the smelting time and increasing productivity. However, such solution can cause some scrap oxidation, "jamming" of scrap in the basket, reduction of the basket life, and additional logistic or operational difficulties. Experiences show that for smaller furnaces with low power transformers, such a solution can bring economic effects, while for large furnaces with high-power transformers, the benefits of electricity savings and other effects are relatively small and usually such a solution is not economically justified.

Another concept is the scrap heating system in the shaft. In this case the structure of the furnace vessel, especially the roof, must be completely changed. An example

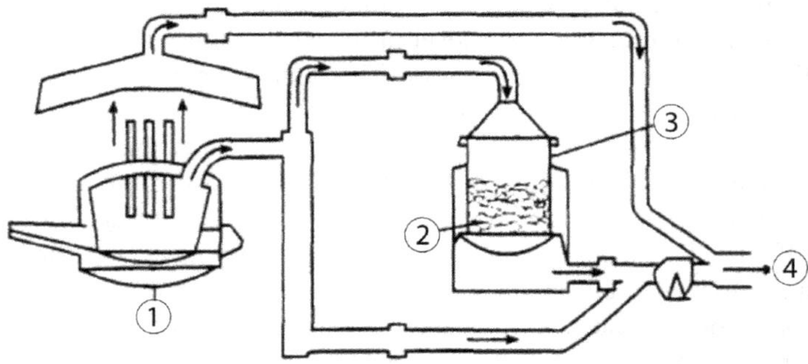

FIGURE 7.12 General concept scheme of the system of scrap heating up with waste gases in the charging basket: 1 – electric arc furnace, 2 – charge materials, 3 – loading basket, 4 – off-gas capture system. (Author's work.)

of such a solution is the system shown in Figure 7.13. The furnace vessel has a slightly larger diameter than the classic one, and a shaft-shape steel structure with movable supporting trusses (fingers) from the bottom is installed instead of the roof elbow. The waste gases from the metallurgical process directly rise to the shaft and are then transferred to the off-gas system. Cold scrap is placed in the shaft and is heated by the energy contained in the hot flowing gases. The heated scrap from the shaft is fed into the vessel by moving the truss (fingers) away, while the cold scrap is fed from the top to the shaft. The scrap may be heated partly by the waste gas and partly by additional burners placed in the side walls.

It can be assumed that the temperature of gases fed into the chamber shaft is practically equal to the temperature of gases taken from the vessel, i.e. about 1,873 K (1,600°C), and the temperature of those leaving the shaft is about 1,273 K (1,000°C). The degree of heating up depends on the type of scrap, and it can be expected that the temperature of scrap after heating up will be between 1,073 K (800°C) and 873 K (600°C). This will enable savings of electricity by about 100–120 kWh/ton of steel, reduction of electrode consumption by about 0.4 kg/t, as well as shortening the smelting time and increasing efficiency. However, such a solution brings many logistic inconveniences related to the loading of scrap, its backfilling from the shaft to the vessel, durability of equipment installed in the shaft, operating at high temperature, etc. On the one hand, pilot installations of this type show the energy effects of the solution; on the other hand, the problems with the operation of the heating shaft and some limitations in the flexible conduct of the smelting technology do not recommend this type of construction for common, industrial implementation.

Structures of scrap heating systems are also being developed, similar in concept to those above, but using a horizontal shaft. A general scheme of such a system is shown in Figure 7.14. In such a solution, in the furnace vessel, an opening is made in the

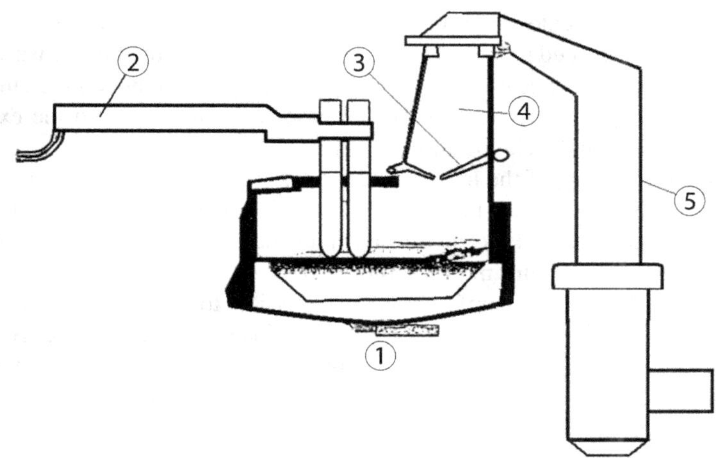

FIGURE 7.13 General scheme of the furnace with a shaft for heating scrap before loading into the furnace vessel: 1 – electric arc furnace, 2 – electrode support arms, 3 – fingers supporting the scrap, 4 – scrap heating shaft, 5 – off-gas capture system. (Author's work.)

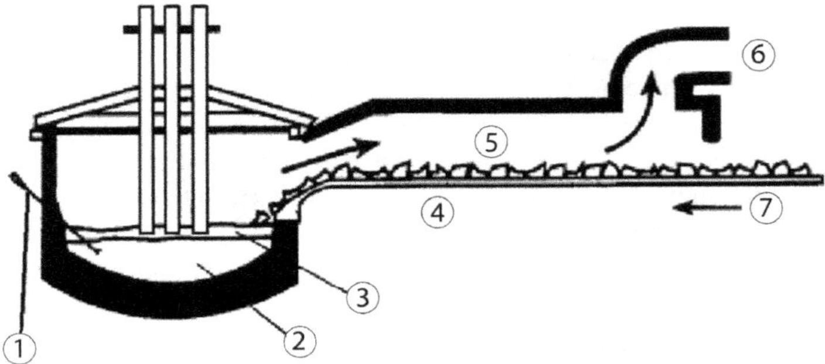

FIGURE 7.14 General scheme of the scrap heating system in the horizontal shaft: 1 – oxygen lance, 2 – metal bath, 3 – foaming slag, 4 – electric arc furnace, 2 – electrode support arms, 3 – fingers supporting the scrap, 4 – heater, 5 – off-gases, 6 – for cleaning off-gases from dust, 7 – direction of the scrap movement. (Author's work.)

walls for waste gas pulling out and scrap loading. The horizontal shaft is equipped with a conveyor belt for loading scrap. The loaded cold scrap on the conveyor slowly moves toward the furnace vessel, while the hot waste gases get carried away in the opposite direction. As a result, heat energy from gases is transferred to scrap.

An example of the practical concept of such a system is proposed in the solution called Consteel® [17]. The proposed solution involves the process of continuous loading of scrap onto a conveyor belt from a landfill or directly from railroad wagons, using a crane. The slowly moving conveyor moves the scrap toward the furnace and finally fills it into the vessel. Scrap on the conveyor, before it reaches the furnace, enters the horizontal housing of the heating section, where it is heated by hot gases sucked from the vessel. The gases are transferred in the opposite direction to the scrap movement, which means that the gases of the highest temperature, coming directly from the furnace, go to the most heated scrap metal, while the already cooled ones go to the cold scrap. In the scrap heating section, near the furnace, air blowing lances are placed to burn the carbon monoxide contained in the waste gases. This results in additional heat energy, and the exhaust gases do not contain unwanted CO. After leaving the horizontal shaft, the gases are transferred to the extraction system similarly as in the classic case.

Continuous charging of the heated scrap to the furnace takes place into an existing already metal bath, thanks to which it has the opportunity to dissolve after being immersed in liquid metal. Electric arcs burn all the time over the liquid metal, not over solid scrap, so they burn more stable.

The Consteel® solution, therefore, makes it possible to carry out both scrap heating using the heat of waste gases and continuous loading of the charge. Both technologies together bring many benefits, especially by reducing production costs, increasing production efficiency, increasing production flexibility, and improving environmental aspects. Currently, there are several dozen such installations in operation worldwide. Confirmed effects in electricity savings depending on the operating conditions are from 80 to 120 kWh/ton of steel, with an average temperature of scrap heating in the range of 673–873 K (400–600°C).

REFERENCES

1. Jensen J., Wolf K.: Reduction of EAF Dust Emissions by Injecting It into the Furnace, *Metallurgical Plant and Technology*, Vol. 3, 1997, pp. 58–62.
2. Puntigam A., Rotsch J.: Trends und Entwicklungen der Feuerfesttechnik im Bereich der E-Ofen und Konverter Stahlerzeugung, Tagungsband, *125 Jahre Institut fur Eisen- und Stahltechnologie*, 1999, 4–5 Oktober, Freiberg (in German).
3. Jones J.A.: New Steel Melting Technologies, *Iron and Steelmaker*, Vol. 9, 1997, pp. 69–71.
4. Jones J.A.: New Steel Melting Technologies, *Iron and Steelmaker*, Vol. 10, 1997, pp. 109–112.
5. Sofilic T., Rastovcan-Mioc A., Smit Z.: Polychlorinated Dibenzo-*p*-Dioxin and Dibenzofuran Emissions from Croatian Metallurgical Industry, *Archives of Metallurgy and Materials*, Vol. 53, 2008, No. 2, pp. 583–594.
6. Birat J.-P., Arion A., Faral M., Baronnet F., Marquaire P.-M., Rambaud P.: Abatement of Organic Emissions in EAF Exhaust Flue Gas, *La Revue de Metallurgie-CIT*, Vol. 98, 2001, pp. 839–854.
7. J. Jablonski: Technologie zero emisji, Wydawnictwo Politechniki Poznanskiej, in *rozdział IV: Problematyka odpadow w przemysle metalurgicznym*; M. Karbowniczek, Z. Wcislo Editors, 2011. ISBN 978-83-7775-063-6 (in Polish).
8. Niesler M.: Najlepsze dostepne techniki (BAT). Wytyczne dla produkcji stali – stalownie elektryczne z odlewaniem stali, Ministerstwo Srodowiska, 2005 (in Polish).
9. Niesler M.: Najlepsze dostepne techniki (BAT). Wytyczne dla produkcji zelaza i stali – huty zintegrowane, Ministerstwo Srodowiska, 2005 (in Polish).
10. Karbowniczek M., Kawalkowski M.: Wykorzystanie CFD do modelowania ukladu odciagu gazow z elektrycznego pieca lukowego, *Informatyka w Technologii Materiałów*, Vol. 5, 2005, No. 3, s.61–s.73 (in Polish).
11. Karbowniczek M., Kawalkowski M.: Model układu odciagu gazow z elektrycznego pieca lukowego, Konferencja, *Teorie a praxe wyroby a zapracowani oceli*, Roznov (Czech Republic), 4–5 April 2006, s.157–s.164 (in Polish).
12. Karbowniczek M., Kawalkowski M.: The electric arc furnace off-gasses modeling using CFD, *16th Steelmaking Conference*, 5–8. 11.2007, Rosario, Argentina, pp. 127–133.
13. Unamuno I., Laraudogoitia J., Almeida S.: Sidenor Basauri EAF Emissions Reduction through Analysis and Modelling, *Archives of Metallurgy and Materials*, Vol. 53, 2008, No. 2, pp. 379–384.
14. United States Environmental Protection Agency, National Emission Standards for Hazardous Air Pollutants (NESHAP) for Iron and Steel Foundries – Background Information for Proposed Standards, 2002.
15. Deppner K.H.: Verticon – An Efficient and Environmentally Friendly Scrap Preheater, *Steel Times International*, Vol. 11, 1998, pp. 20–23.
16. Born C., Granderath R.: Potential and Difficulties of Heat Recovery in Steel, *Metallurgical Plant and Technology International*, Vol. 4, 2013, pp. 50–60.
17. Varini R., Ferri M., Raggio C.: Tenova iron & steel raw materials flexibility and green environment: Key success factors in electric steelmaking, *Proceedings of 10th European Electric Steelmaking Conference*, Graz, Austria, 2012.

8 Charge Materials

Charge materials used for steelmaking processes can be traditionally divided into three basic groups:

1. Metallic materials:
   - steel scrap,
   - direct reduced iron (DRI),
   - crude iron (pig iron),
   - ferroalloys and technical metals.
2. Slag-forming materials:
   - metallurgical lime,
   - dolomite,
   - bauxite.
3. Oxidizing and carburizing materials:
   - oxygen gas,
   - coal materials for carburization and slag foaming.

8.1 STEEL SCRAP

Steel scrap is the basic charge material for the production of steel in an EAF. This is the main material that brings iron to the melting process [1]. The criteria for classifying steel scrap by types and a description of the scrap quality are laid down in many national standards, governing the principles of scrap trade. The provisions of the standards usually break scrap down into:

1. *Types*: depending on the degree of preparation for the EAF application, the following are distinguished:
   W – prepared non-alloy scrap,
   S – prepared alloy scrap,
   N – unprepared non-alloy scrap,
   NS – unprepared alloy scrap.
2. *Classes*: depending on the form, dimensions, density, and degree of corrosion and acceptable metallic and nonmetallic impurities, scrap is broken down into classes denoted with numbers, for the prepared and unprepared scrap, respectively.
3. *Categories*: depending on the number and type of alloy-forming elements, the prepared scrap is divided into the following categories:
   unprepared scrap – no categories
   A – non-alloy steel scrap with a limited content of some elements,
   B – alloy steel scrap containing manganese and/or silicon,
   C – alloy steel scrap containing chromium,

DOI: 10.1201/9781003130949-8

D – alloy steel scrap containing chromium and molybdenum,
E – alloy steel scrap containing chromium and nickel,
F – alloy steel scrap containing chromium, nickel, and molybdenum,
G – alloy steel scrap containing chromium, nickel, and tungsten or molybdenum,
H – alloy steel scrap containing chromium, nickel, molybdenum, and copper,
I – alloy steel scrap containing chromium and tungsten,
K – alloy steel scrap containing chromium, tungsten, and molybdenum,
L – alloy steel scrap containing cobalt,
M – scrapped alloys.

4. *Groups*: Prepared non-alloy scrap of type W constitutes one undivided group. Prepared non-alloy scrap of category A with a limited content of some alloying elements is another group. Prepared alloy steel scrap comprises a group of type S and categories from B through L, and scrapped alloys of category M.
5. *Degree of corrosion*: there are three degrees of prepared scrap corrosion, defined as per Table 8.1.

Standards often also contain requirements concerning preparation for prepared and unprepared scrap. The preparation of scrap, as regards both non-alloy and alloy scrap, involves sorting, shearing, and baling. Sorting the prepared scrap pieces by dimensions and mass is to assign their sizes and types to individual classes. Pressed bales of sheets, wires, and turnings, with their mass of 1 dm^3 under 1 kg, should not be shipped to a customer but should rather be pressed at higher capacity balers.

Standards also contain division of scrap according to its chemical composition. Prepared scrap and unprepared scrap with a limited content of some elements should be sorted by the chemical composition to categories and groups. Non-alloy scrap must not contain alloy scrap, alloy cast iron, nonferrous metals, and their alloys. Classifying plated and bimetal scrap containing more than one steel grade to an appropriate group should depend on the average content of alloying elements in this scrap.

The requirements for the individual classes of prepared non-alloy scrap can be the following, as an example. Scrap of class W1 should be in pieces, in the form of plastic forming or steelmaking process scrap, and it can be post-consumer scrap, except wire and wire products or tubular sections. The maximum sizes of pieces should not exceed 1,000×500×500 mm or a thickness of 6 mm and above; the biggest piece

TABLE 8.1
The Degrees of Prepared Scrap Corrosion (Author's Work)

Degree of Corrosion	Description of Scrap Corrosion
I	Not rusty with a metallic surface or with its surface covered with scale or with rust tarnish
II	Rusty with a surface covered with a rust layer
III	Rusted-through, burned-through, or burned with acids and lyes

mass is 2,000 kg, and protruding parts in the case of structural scrap should not exceed 100 mm. The bulk density of the scrap is approximately 0.8 t/m^3, whereas the acceptable level of metallic and nonmetallic impurities in the II degree of corrosion scrap must meet the following condition: nonmetallic impurities that cannot be separated during scrap processing are allowed up to 2%.

Class W2 contains post-consumer scrap in pieces, tubular sections, coiled or bundled strips, cables, wires, and bars with dimensions up to 1,000×500×500 mm, while the minimum thickness is 4 mm. This class includes strips, wires, ropes, and bars in coils of a diameter up to 600 mm, tubular sections with a wall thickness minimum of 3 mm, and an outer diameter up to 150 mm. The biggest mass of a single scrap piece must not exceed 2,000 kg. The bulk density of this type of scrap is approximately 0.7 t/m^3. Nonmetallic impurities that cannot be separated during scrap processing are allowed in an amount not exceeding 2%.

Class W3 is represented by scrap in pieces without tubular sections, wire and wire products, cables, and turnings. The largest allowed dimensions of this scrap are 300×300×200 mm and a thickness of above 8 mm. The bulk density of the scrap should be approximately 1.3 t/m^3. The content of nonmetallic impurities, which cannot be separated during scrap processing, should not exceed 1.5%.

Class W4 includes special scrap in pieces: uniform without tubular sections, wires and wire products, and turnings. It is the so-called blue scrap and fabrication scrap (e.g. rails, post-production waste derived from, rolling mill, stamping press, shipyard); the largest allowed dimensions are 300×300×200 mm and the thickness of plate pieces 10 mm and more; bar diameter minimum 20 mm; the bulk density of the scrap approximately 1.2 t/m^3. The content of nonmetallic impurities in the chemical composition of this scrap cannot exceed 1%.

Class W5 includes special scrap in pieces: uniform without tubular sections, wires and wire products, and turnings. Similar to W4, it is the "blue" scrap and post-production scrap; the biggest dimensions are 1,000×500×500 mm and the thickness of plate pieces 10 mm and more; bar diameter minimum 20 mm; mass of individual pieces up to 35 kg; bulk density of the scrap approximately 1.2 t/m^3. No impurities should be present in this scrap class.

Class W6 is represented by scrap in pieces without wire sections and wire products, ropes, and turnings. The biggest dimensions are 500×300×200 mm, the thickness of scrap pieces above 6 mm; the biggest mass of an individual piece 35 kg; upon the customer's consent, the mass of individual pieces can be up to 100 kg; scrap bulk density 1.2 t/m^3, the acceptable content of nonmetallic impurities that cannot be separated during scrap processing should not exceed 1.5%.

Class W7 includes scrap in pieces without wires and wire products, whereas tubular sections are allowed. The largest dimensions of this scrap are 1,000×500×500 mm, and the minimum thickness is 8 mm. The mass of a single piece is up to 1,000 kg; scrap bulk density is approximately 1.0 t/m^3. Up to 1.5% nonmetallic impurities that cannot be separated during scrap processing are allowed.

Class W8 is represented by scrap in pieces "shortcut" from a hydraulic press shearing machine without wires and wire products, but tubular sections are allowed. The largest allowed dimensions are 1,000×500×500 mm; scrap bulk density approximately 1.0 t/m^3. Nonmetallic impurities are allowed up to 1.5%.

Class W9 includes mechanically pressed bales from sheets, strips, ropes, wires, and turnings, and scrap in pieces with a thickness adjusted to a specific press. The biggest acceptable dimensions are 2,000×1,000×800 mm; the mass of 1 dm^3 of a bale not less than 2 kg. Up to 3% nonmetallic impurities are allowed; galvanized scrap is allowed up to 2% of the batch mass, and with enamel coatings up to 2% of the batch mass, too.

Scrap class W10 is in the form of mechanically pressed bales from sheets, strips, ropes, wires, and turnings. The largest acceptable dimensions are 600 ×500×300 mm; the mass of 1 dm^3 of a bale not less than 1.0 kg. Scrap with zinc and enamel coatings is not acceptable; nonmetallic impurities are allowed up to 2%.

Class W11 includes mechanically pressed bales from sheets, strips, ropes, wires, and turnings. The largest acceptable dimensions are 1,500×800×800 mm; the mass of 1 dm^3 of a bale not less than 1 kg. Up to 3% nonmetallic impurities are allowed; galvanized scrap is allowed up to 2%, and with enamel coatings up to a 2% batch mass, too.

Class W12 comprises briquettes made from steel turnings, cold and hot pressed. The mass of a single briquette cannot be less than 2 kg; the mass of 1 dm^3 minimum 4 kg; the amount of turnings that chipped in transport must not exceed 5% mass. Nonmetallic impurities should not exceed 3% batch mass.

Class W13 contains loose turnings and small (fine) pieces of scrap. The length of turnings is up to 150 mm; the largest allowed lengths are up to 200 mm, with their amount not exceeding 5% batch mass. Nonmetallic impurities should not exceed 3% batch mass; non-alloyed steel turnings should not be mixed up with other metal, alloyed steel, or cast iron turnings.

Class W14 includes scrap from a foundry, such as runner scrap sprues and cores, center runners, skulls, and others not requiring mechanical processing. The largest allowed dimensions are 1,200×250×250 mm; the largest piece mass is up to 1,000 kg. The allowed amount of nonmetallic impurities should not exceed 5%.

Class W15 includes mechanically processed steel skulls. The largest acceptable dimension is 800 mm, and the maximum mass of one piece is up to 1,500 kg. Processed steel skulls should not contain more slag impurities than 10%.

Class W16 contains blast furnace scrap, including scrap in pieces, rusty, and baled steel turnings mixed with cast iron turnings. The biggest acceptable dimensions are 150×150×150 mm. The scrap can be rusty, burned-through, "burned" with acids and lyes, with or without enamel, paint, varnish, or tar coatings. It can contain other metal coatings except lead, and scrap contamination with zinc or tin should not exceed 0.5%. The total nonmetallic impurities should not exceed 1.5% scrap batch mass.

Class W17 includes steel cables of various lengths, with cut ends (free of nonferrous metals), without specifying lengths, supplied in tipping wagons. Nonmetallic impurities up to 3% are allowed.

Classes W18 and W19 include shredded scrap in pieces. In addition, in class W18 pieces with dimensions 250×250×250 mm and bulk density not less than 1.0 t/m^3 are allowed, and acceptable impurities must not exceed 1.5%. However, in class W19 pieces with dimensions 150×200×150 mm and bulk density not less than 1.2 t/m^3 are allowed, and acceptable impurities must not exceed 1%.

We can assume, for instance, that the scrap of classes W4, W5, W18, and W19 has I degree of corrosion, class W16 has III degree of corrosion, whereas other classes have either I or II degree of corrosion. The criteria and quality of steel scrap contained in the provisions of standards are usually general and are applied in commercial relationships between suppliers and steelmaking shops. In individual cases, arrangements concerning scrap trading that are not covered by the standard criteria may be made. From the point of view of the scrap quality, in particular as regards the content of undesired components, the scrap price is an important aspect.

Steel scrap can be classified by its sources of origin as post-production (new) scrap and so-called post-consumer (old) scrap. The post-consumer scrap contains end-of-life and damaged objects, structures, machinery, and products. It is sourced from industrial plants, construction and erection companies, transport, households, and farms. It is characterized by high diversity, and it can contain all of the steel grades, alloyed, non-alloyed, nonferrous metals, and nonmetallic impurities. Its chemical composition is usually unknown. Post-production scrap is generated in the process of steel manufacturing, its plastic forming, and machining. Steel plant departments (steelmaking shops, foundries, rolling mills, forging shops, and others), machining workshops, steel structure manufacturers, etc., are its sources. Post-production scrap that is generated in steel mills is called internal or home scrap (returns). Usually its chemical composition is known, which facilitates its utilization.

Scrap that is used by a steelmaking shop as a charge must be properly prepared prior to loading into a furnace. This preparation is partially carried out in companies dealing with scrap collection and preparation for sale, whereas the final preparation takes place in the steelmaking shop. Preparation mainly includes segregation into individual types and classes of scrap, breakdown by the content of unwanted elements, and breakdown by the form (so-called high-, medium-, and low-density scrap). In the charging bays of steelmaking shops, the individual types and classes of scrap are placed in separate bins so that it can easily be loaded to a charging basket.

The so-called collection scrap is one of the types of post-consumer scrap. It is collected at purchasing centers, comes from various sources, and is characterized by very diversified chemical compositions and forms; it is often contaminated with nonferrous metals and nonmetallic elements. Preliminary segregation and treatment of scrap like this is performed in purchasing centers. The initially prepared scrap is then transferred to preparation plants, where further segregation and so-called baling is carried out with special machinery called balers, which mechanically compact the spatial scrap into the form of compacted bales with set dimensions.

Companies utilizing end-of-use cars, white goods (refrigerators, washing machines), or other similar equipment are a separate group of scrap preparing plants. In companies like these, scrap is prepared by dismantling individual components and dividing them by material types, e.g. separately copper electric wires, rubber, plastics, steel components, etc. Methods of hydraulic pressing the whole piece of equipment are also applied, e.g. a car with the battery and fuel removed. The pressed equipment is then cut into fine fragments, which are subjected to segregation into individual materials. Operations of this type are carried out in facilities called press shearing machinery.

8.2 DIRECT REDUCED IRON (DRI)

Apart from scrap, as the fundamental charge material, more and more often the so-called DRI, which is also called sponge iron, is used in the steel production process in an electric arc furnace (EAF). This material (in lump, granulated, briquetted, or powder form) comes from the direct reduction of iron ores with a reducing gas. A mixture of hydrogen and carbon oxide is used as the reducing gas. It is made of natural gas and a product of coal (or coke) combustion. The ores are reduced in the solid phase, within the temperature range of 1,073–1,273 K (800–1,000°C). The production process can be carried out in rotary, shaft, or fluidized bed furnaces. The DRI production in the world has currently reached a level of almost 80 million Mg. Equipment from Midrex [2] has the biggest share in this production. In the production practice, various processes based on a shaft or rotary furnace are applied. The other less significant supplier of equipment and practices for the DRI production is Tenova HYL [3]. This company employs a practice based on a shaft furnace.

Three basic varieties of materials are produced by direct reduction:

- cold direct reduction iron (CDRI), that is, DRI in cold form,
- hot direct reduction iron, that is, DRI in hot form,
- hot briquetted iron (HBI), that is, DRI in the form of briquettes.

Regardless of the variety, these materials feature the following parameters: degree of metallization, content of carbon, gangue, and other undesired impurities. The chemical parameters of the materials produced, regardless of their variety, are shown in Table 8.2. The direct reduced materials that are made in the form of briquettes (HBI) and cold pellets (CDRI) are presented in Figure 8.1.

Materials from the direct reduction of iron ores are applied instead of scrap, as an iron-bearing substitute. Steel production practices based on scrap as a charge material are the most economically viable, including from the perspective of environmental protection. However, it is not possible to satisfy the global demand for steel worldwide with scrap only, which is caused by its short supply. Although the whole

TABLE 8.2
Chemical Parameters of Direct Reduction Materials (Author's Work)

Parameter	Value (%)
Degree of metallization	90–95
SiO_2	1.0–3.0
Al_2O_3	0.5–2.5
CaO	0.1–1.5
MgO	0.1–1.0
$Na_2O + K_2O + TiO_2$	Max. 0.15
S	Max. 0.01
C	0.5–2.5

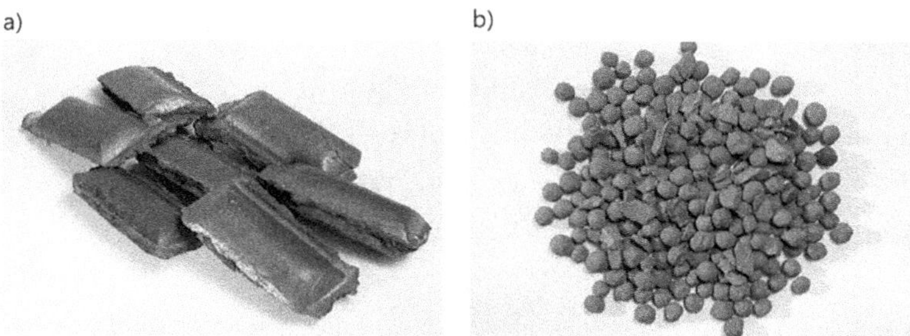

FIGURE 8.1 View of direct reduced materials in the form of: (a) HBI and (b) CDRI. (Author's photo.)

mass of scrap generated worldwide is used in the production process, it can cover less than half of the demand. Production practices based on the BOF process, using hot metal produced in the blast furnace process, are less economically and environmentally justified. This led to efforts to find a steel production practice using iron ores, while omitting the blast furnace process. DRI materials are such an attempt. However, note the fact that replacing scrap with DRI in the EAF involves changes in the charge loading practice, its meltdown, and slag formation [4].

8.3 CRUDE IRON

Crude iron is used in steelmaking processes, both in the solid form (pig iron) in EAF s and in the liquid form (hot metal) in basic oxygen furnaces. The chemical composition of crude pig iron is essential for the practice and economy of steel production. It is conditioned by, among others, the production schedule and technical conditions of the steelmaking shop and the capabilities of the blast furnace process.

The quality and technical commercial criteria of pig irons are adjusted to the technical capabilities of blast furnaces and steelmaking shops. Pig iron is one of the types of ferrous alloys which feature a relatively high carbon content (it is accepted that above 2%). There are no clear cut limits defining the carbon content in iron. In addition, these limits are usually not specified in the standards. In practice, the carbon content is usually from 3.8% to 4.5%. The limits of other constituents of iron are given in Table 8.3.

The acceptable contents of the individual constituents specified in the standard are relatively high, but for the steelmaking process, they must be much lower. As a rule, in individual commercial deliveries, the chemical composition of iron is more restricted. An example of a restricted chemical composition of pig iron is given in Table 8.4. This is restricted due to the process and economic considerations of the steelmaking shop production process.

In the conditions of a steelmaking process, silicon and partially also manganese easily undergo exothermic oxidation reactions, emitting significant amounts of heat, which is beneficial from the perspective of energy. However, these reactions result in the formation of oxides, which increase the amount of the forming slag, and the forming SiO_2 also requires adding increased amounts of lime to ensure the

TABLE 8.3
General Limits of Pig Iron Constituents (Author's Work)

Constituent	Value (%)
Mn	≤30
Si	≤8
P	≤3
Cr	≤10
Other total	≤10

TABLE 8.4
Example of Limits of the Chemical Composition of Iron for Steelmaking Purposes (Author's Work)

Constituent	Value (%)
Mn	≤0.9
Si	≤0.6
P	≤0.1
S	≤0.03
other total	≤1.0

proper slag basicity. It affects the life of the refractory lining of the walls and hearth. Phosphorus and sulfur are detrimental impurities in steel; therefore, it is practically always necessary to reduce their content in a steelmaking process to an as low as possible level. Therefore, it is vital to use charge materials, including pig iron, with the lowest phosphorus and sulfur contents as possible.

Crude iron in electric steelmaking shops is as a rule applied in the form of ingots (so-called pigs), and the weight of individual pigs ranges from 20 to 50 kg. Other materials with similar physicochemical properties, such as iron scrap (broken ingot molds) and iron skulls, can be included in the iron group. Figure 8.2 offers the view of pig iron.

Pig iron is used in an EAF as an additional material bringing iron. It can be applied as a carbon carrier to the charge or a material dissolving unwanted elements in the melt (e.g. copper, tin, lead, etc.). From the practice perspective, pig iron is an advantageous material in an EAF, but the decision about its application is made on business grounds and depends on price relationships with regard to other carbon-bringing materials, or the costs of removing unwanted components from scrap.

It follows from the literature that, in very rare cases, crude iron as a hot metal can be or is the basic charge material for steel production in an EAF. However, operating experience from a few plants of this type did not show clear benefits of such a

FIGURE 8.2 View of crude iron in the form of pigs. (Author's photo.)

solution. There were some attempts to use hot metal with its various shares in the charge, from 30% to 100% in relation to cold scrap or DRI [4]. However, a practice like this requires quite a different technique during melting and oxidation of the metal bath, as well as the EAF design with a significantly altered vessel shape and the extension of the off-gas evacuation system. At present, such solutions are practically irrelevant on the industrial scale.

8.4 FERROALLOYS AND TECHNICAL METALS

Ferroalloys are alloys of iron with other elements. They are manufactured as indirect (for further processing) materials and practically only applied in the steelmaking industry. They contain some amounts of iron and one or more elements belonging to nonferrous metals or semi-metals, which constitute alloying constituents. Contrary to the name, usually the iron content is lower than that of the other constituents of a ferroalloy. Generally, ferroalloys can be classified by their application in the steelmaking process in an EAF as deoxidizers, alloy additions, and inoculants. The application of ferroalloys is the easiest and most cost-effective method for introducing an alloying element to the melt. At the same time, the production of a ferroalloy on an industrial scale is usually cheaper than producing a material with a high content of any specific element. Generally, ferroalloys should contain as much of the basic constituent as possible, a small amount of unwanted impurities, and an appropriate lump size. The quality and commercial criteria for individual ferroalloys are included in the relevant standards. Following is a short description of the most common ferroalloys that are used in the EAF steelmaking process.

Ferrosilicon: applied as a deoxidizer and an alloying addition for manufacturing high silicon steels, and steels with an increased silicon content. A few ferrosilicon grades are manufactured, including the most often used grades FeSi45 and FeSi75 (the numbers mean the average silicon content in the alloy). Ferrosilicon is made in electric arc-resistance furnaces by the reduction of silica. It is used in the form of lumps. Figure 8.3 offers the view of lump FeSi75.

FIGURE 8.3 View of lump ferrosilicon (75%). (Author's photo.)

FIGURE 8.4 View of lump ferromanganese (70%). (Author's photo.)

Example of the chemical composition restrictions for FeSi75: **C**-max. 0.2%, **Si**-min. 72%, **P**-max. 0.5%, **S**-max 0.01%, **Al**-max. 2.0%.

Ferromanganese and technical manganese: applied as a deoxidizer and an alloying addition for manufacturing high manganese steels, and steels with an increased manganese content. A few ferromanganese grades are manufactured, characteristically with a content of manganese and carbon. Ferromanganese is manufactured in a blast furnace or in an electric arc-resistance furnace. Cheaper high-carbon grades are made in a blast furnace (which is a feature of the blast furnace process), containing 6%–8% C and 60%–70% Mn. Low-carbon (more expensive) grades are produced by the ladle refining of high-carbon FeMn or in the electrothermal process. Technical manganese is produced with electrothermal methods, but due to its high price, it is only applied in special cases. They are all used in the form of lumps. Figure 8.4 offers the view of lump FeMn70.

Example of a chemical composition limitation for FeMn70: **C**-max. 8.0%, **Mn**-min. 68%, **Si**-max. 2.0%, **P**-max. 0.15%, **S**-max. 0.03%.

Ferrochromium and technical chromium: only applied as an alloying addition for manufacturing high chromium steels and steels with an increased chromium content. A few grades of ferrochromium are produced, and the most important criterion for their classification is the carbon content, which determines its price. Groups of high-carbon ferrochromium (HC-FeCr), medium-chromium (MC-FeCr), and low-chromium (LC-FeCr) are distinguished. High-carbon ferrochromium is produced in electric resistance-arc furnaces by the reduction of chromium ores with carbon reducers. Medium- and low-FeCr are produced by refining high-carbon alloys, including primarily vacuum decarburization. HC-FeCr alloys contain 7%–9% C; MC-FeCr alloys contain 3%–5% C, whereas LC-FeCr contain under 0.3% C. All ferrochromium varieties contain 60%–70% Cr. The application of individual FeCr grades depends on the type of steel manufactured. The high-carbon variety tends to be used for manufacturing steel containing up to a few percent of chromium and a carbon content above 0.2%, e.g. tool steel grades. Medium- and low-carbon varieties are primarily used for the production of low-carbon stainless steels (austenitic, martensitic, and ferritic) made with vacuum methods. FeCr is also used in the form of lumps. Figure 8.5 offers the view of lump HC-FeCr.

Example of chemical composition restrictions for HC-FeCr: **C**-max. 10%, **Cr**-min. 60%, **Si**-max. 2%, **P**-max. 0.03%, **S**-max. 0.05%. Whereas for LC-FeCr: **C**-max. 0.3%, **Cr**-min. 60%, **Si**-max. 2.0%, **P**-max. 0.03%, **S**-max. 0.03%, **S**-max. 0.3%.

Silicon-manganese: it is a frequently used three-component ferroalloy, applied as a deoxidizer and an alloying addition for manufacturing steels with an increased content of silicon and manganese at the same time. A few grades of silicon-manganese are produced, but grades containing a dozen or so percent of silicon and 60%–70% manganese are the most often used. FeSiMn is produced in electric resistance-arc furnaces by the reduction of manganese ores containing some amounts of silica. It is used in the form of lumps. Figure 8.6 offers the view of lump FeSiMn.

Example of the chemical composition restrictions for FeSiMn: **C**-max. 1.8%, **Si**-min. 17%, **Mn**-min. 60%, **P**-max. 0.10%, **S**-max. 0.03%.

Aluminum: it is a metal commonly used for the deoxidation of the metal bath and to control the austenite grain size. It can be an alloying addition for manufacturing steels for nitride case hardening and heat-resisting steels. For the manufacturing needs, aluminum is made, depending on the intended use, in the form of ingots (sows), granulate, or wire. Aluminum is a so-called light metal. Its specific gravity is more than three times lower than the specific gravity of iron, which causes

FIGURE 8.5 View of lump high-carbon ferrochromium. (Author's photo.)

FIGURE 8.6 View of lump silicon-manganese. (Author's photo.)

difficulties related to introducing it into the metal bath, involving the needs to apply special methods. When a piece of aluminum is simply dropped onto the melt surface, it floats on the surface and oxidizes, while not dissolving in the metal bath. Therefore, methods of mechanical immersion or introducing to the liquid steel in the form of a wire, with the immersion method, are applied. A few grades of technical aluminum are produced, and the most important parameter is the so-called purity, or the aluminum content in the alloy. Alloys marked with symbols from Al80 to Al97 are distinguished. The number means the percentage share of aluminum. The most important impurities include silicon, magnesium, copper, and zinc. Aluminum is produced from bauxites with hydrometallurgical methods or in electric induction furnaces from aluminum alloy scrap.

Apart from the above-mentioned ferroalloys, dozens of other ferroalloys are applied in the steelmaking practice. They are only used to make up the chemical composition of the metal bath. Binary alloys are used, containing iron and a component being the element to make up the chemical composition, for instance, ferromolybdenum, ferrovanadium, ferrotungsten, ferrotitanium, etc. There are also multicomponent ferroalloys for special purposes, for instance, to bring small amounts of elements to modify the grain size of produced steel. This group includes iron-calcium-silicon, iron-calcium-magnesium, etc. Apart from the above-described aluminum, other technical metals are also applied, in particular to make up the chemical composition of steel, for instance, nickel or cobalt.

8.5 SLAG-FORMING MATERIALS

Slag-forming materials are used to produce slag in the steelmaking process – an ionic solution with a nonmetallic nature, with the optimum chemical composition and an adequate amount. Metallurgical lime is the basic slag-forming material in the EAF, including its variety, the so-called dolomite metallurgical lime. In addition, in some cases, burned dolomite can be used, calcined bauxite and, in special cases, other materials.

Metallurgical lime, being a CaO carrier in the slag, is produced from a mineral called limestone. The production process includes burning (calcining) in lime kilns, primarily of the shaft type, within the temperature range of 1,273–1,473 K (1,000–1,200°C). The following parameters, defining the metallurgical lime quality, determine its suitability for metallurgical processes:

- chemical composition,
- lump size,
- reactivity,
- porosity,
- time from the burning to the application in the metallurgical process.

The criteria for individual parameters are described in the standards. There are a few lime grades, used in steelmaking processes. In an EAF, high reactive burned lime in various size lumps is applied. In the chemical composition, the CaO content is specified, which should be a minimum of 92%–94%, and the allowed MgO content, usually a maximum of 0.6%. In addition, the so-called active CaO plus MgO sum content is specified, which should be a minimum of 90%–92%. Lime is produced with various lump sizes, adjusted to the method of introducing it to the metal bath. For instance, if it is be loaded to a charging basket, the applied lump size is 20–50 mm; for injection to the metal bath, the size applied is from 0.5 to 6 mm, or some other size subject to the needs.

Sometimes, dolomite lime is applied, where the MgO content is from 2% to 12%, while the other parameters are analogous to the "regular" lime. Figure 8.7 offers the view of lump burned lime.

Burned dolomite is used in steelmaking processes as the MgO carrier in the slag. It is a product of burning a mineral called dolomite at a temperature of approximately 1,173 K (900°C). Chemically, dolomite is a mixture of lime and magnesium carbonates. The purpose of the burning process is to evaporate moisture, and most of all to transform carbonates into the oxides of these elements. During the burning, 40%–50% of the mass of natural dolomite is released in the form of carbon dioxide. Depending on the type of the natural dolomite bed, the chemical composition of burned dolomite, applied in metallurgy, is as follows: 30%–40% CaO, 20%–50% MgO, and up to a few percent of impurities like SiO_2, Al_2O_3, and Fe_2O_3. In industrial practice, dolomite with a lump size from 2 to 20 mm is applied. Figure 8.8 offers the view of lump burnt dolomite.

FIGURE 8.7 View of lump burnt lime. (Author's photo.)

FIGURE 8.8 View of lump burnt dolomite. (Author's photo.)

FIGURE 8.9 View of lump calcined bauxite. (Author's photo.)

Bauxite is the carrier of Al_2O_3 in the slag in steelmaking processes. It is usually used calcined. Bauxite is a sedimentary rock consisting mainly of aluminum hydroxides ($Al_2O_3.nH_2O$), ferrous hydroxides ($Fe_2O_3.nH_2O$), silica, and titanium oxide. Bauxite calcination is a variety of roasting, where it is heated to a temperature ranging from 1,473 to 1,623 K (1,200–1,250°C), to decompose aluminum hydroxide, and to remove the water from its crystal structure. After calcination, aluminum oxide (Al_2O_3) is obtained. Typical technical calcined bauxite contains: min. 85% Al_2O_3, max. 6% SiO_2, and max. 2% Fe_2O_3. In industrial practice, calcined bauxite with a grain size from 5 to 8 mm is applied. Figure 8.9 offers the view of lump calcined bauxite.

Calcium carbide CaC_2 is one of the materials used to control the slag chemistry in the EAF. Typical technical calcium carbide contains: approximately 70% CaC_2, approximately 27% CaO, and up to 3% sum (MgO, Al_2O_3, FeO, and SiO_2). In industrial practice, calcium carbide sized 15–25 mm is applied. Figure 8.10 offers the view of lump calcium carbide.

8.6 OXIDIZING AND CARBURIZING MATERIALS

In the steelmaking processes carried out now, oxygen in the gaseous form is the basic oxidizing material, or the material supplying oxygen to the metal bath. In the EAF, the technology of supplying oxygen with lances made of consumable steel pipes or water-cooled lances is applied.

FIGURE 8.10 View of lump calcium carbide. (Author's photo.)

Carbon has various functions in the steelmaking process:

- alloying component in steel,
- material providing (increasing) the carbon content in the metal bath in order to carry out the oxidation period,
- expected source of thermal energy coming from its oxidation,
- slag foaming agent.

In industrial practice, various materials containing carbon as the basic element are applied. These materials, due to their function in the process, are called "carburizing agents" or "foaming agents". Carburizing agents are used to increase the carbon content in the metal bath. There are two types of carburizing agents, differing with the lump size:

- with a lump size from 25 to 70 mm fed to the charging basket together with scrap,
- with a lump size from 1 to 3 mm fed to the metal bath by injection.

The foaming agents are used to generate foamy slag, and they are introduced to the slag volume by injection. In this case, the grain size should be from 0.5 to 2 mm [5].

From the point of view of chemistry, all types of carbon bringing materials should meet similar requirements:

- elementary carbon content as high as possible (min. 85%),
- sulfur content as low as possible (max. 1%),
- ash content as low as possible (max. 12%),
- volatile matter content as low as possible (max. 3%),
- as little moisture as possible (max. 3%).

Materials bringing carbon in the steelmaking process are prepared from various input materials: various hard coal grades (including anthracite, for example), coke, graphite (usually as post-production scrap or end-of-life graphite products, e.g. graphite electrodes), other materials meeting the chemistry requirements, and coming from the post-production processes of the chemical industry, etc. Suppliers of these materials sometimes use blends of a few various input materials, according

FIGURE 8.11 View of anthracite carburizing agent with a grain size from 1 to 3 mm. (Author's photo.)

FIGURE 8.12 View of anthracite carburizing agent with a grain size from 25 to 70 mm (Author's photo.)

to strictly selected recipes. Before blending, the input materials are crushed to the appropriate fractions and often dried.

Figures 8.11 and 8.12 offer the view of examples of carbon bringing materials in steelmaking processes in an EAF in lump form.

REFERENCES

1. Tardy P.: Demand and Arising of Ferrous Scrap: Strains and Consequence, *Archives of Metallurgy and Materials*, Vol. 53, 2008, No. 2, pp. 337–343.
2. The MIDREX® Process: The World's Most Reliable and Productive Direct Reduction Technology, 2014, corporate brochure.
3. Becerra J., Martinis A.: Practical Options for Advancing the Direct Reduction Industry in India, Millenium Steel, 31 October 2008.
4. Hornby S., Madias J., Torre F.: Myths and realities of charging DRI/HBI in electric arc furnaces, *AISTech 2015 Proceedings*, Cleveland, OH, pp. 1895–1905.
5. Karbowniczek M., Wolanska E., Mroz J.: Wpływ rodzaju materialu spieniajacego na parametry pienienia zuzla w piecu lukowym, *Hutnik Wiadomosci Hutnicze*, Vol. 9, 2004, pp. 464–468 (in Polish).

9 Steel Production Technique in Arc Furnaces

The process of steel production in electric arc furnaces (EAFs) is based on scrap as the basic charge material [1]. Other materials are also used, which are carriers of iron in the charge, but the share in world steel production of these charge materials is small. They include materials produced from iron ore, using the direct reduction practice, sometimes crude iron. On the one hand, the type of material used does not determine the course of steel melting in an EAF, except for some differences in the method of melting down. On the other hand, the chemical composition of the scrap processed has a significant influence on the melting practice. When processing scrap in EAFs with basic refractory lining, three fundamental groups of practices are applied:

- non-alloy charge (scrap) processing with its simultaneous melting and oxidizing with oxygen gas, and the obtained semi-product is then refined in a ladle furnace (LF) or other device for the so-called secondary metallurgy,
- processing of the alloy charge (scrap) with the use of recovery methods, without oxidizing with oxygen gas or with limited refining, and the obtained semi-product is then refined in vacuum secondary metallurgy facilities,
- processing of high-chromium charge (scrap) using recovery methods or with limited oxidizing with oxygen gas, and the obtained semi-product is then refined in vacuum secondary metallurgy facilities.

The first group of methods is used for the melting structural carbon and low-alloy steels, using non-alloy or low-alloy scrap or materials derived from the direct reduction of iron ore as the charge. Practices like these are rather used in steelmaking shops producing large quantities of steel, equipped with a large furnace, with an installed high-power transformer and using additional facilities to intensify the process (burners, lances, etc.). A semi-product (metal bath) is produced in the furnace. After being tapped into a ladle, it is subjected to operations of refining and obtaining finished steel in secondary (ladle) metallurgy facilities. The process comprises two production stages. The first stage is carried out in the furnace, and its objective is to obtain a metal bath with the minimum content of alloy elements in as fast and cost-effective manner as possible, using one slag. The second stage is usually carried out in a ladle furnace, where refining operations are conducted, most often under two slags, including deoxidation, replenishment of chemical composition, and heating to a temperature suitable for casting the obtained steel.

DOI: 10.1201/9781003130949-9

The second group of practices is used for melting medium, and in particular high-alloy steel grades, using alloyed scrap as the charge. Practices like these are used rather in steel plants producing small quantities of steel, equipped with a small furnace, with a lower power transformer installed, and only some equipment to intensify the operation. The furnace can produce, as before, a semi-product, which after being tapped into a ladle is subjected to operations of refining, and obtaining finished steel in secondary metallurgy facilities, often using vacuum. This group also distinguishes single-stage practices, in which the entire production process, including melting and refining, is carried out in the furnace, under two slags. The finished steel for casting is then produced directly in the furnace. The lack of oxidation period or limited oxidation is a characteristic feature of these methods, and the primary purpose is to recover valuable alloying elements from scrap.

The third group of methods is used to melt the so-called high-chromium steels, i.e. containing at least a few percent of chromium in their chemical composition. This especially applies to martensitic, ferritic, or austenitic steel grades, which feature a chromium content in the range of 15%–30% and carbon in the range of 0.02%–0.08%. Due to high prices of chromium as an alloying element, it is beneficial to produce steels of this type using scrap of these steels as the charge material. At the same time, in the steelmaking process there is the need to oxidize carbon to a low level. In the usual conditions of the steelmaking process, the oxidation of carbon is associated with the simultaneous oxidation of chromium, i.e. the oxidation of carbon will cause significant losses of chromium recovered from the scrap. The solution to the above problem is to conduct the steelmaking process at very high temperatures (from 2,023 to 2,123 K (1,750°C–1,850°C)) or to conduct the process of carbon oxidation in the vacuum. It is only in conditions like this that carbon is oxidized without oxidizing chromium. At present, this type of steel is manufactured in process lines including an EAF and ladle steel refining facilities. The most frequently used facilities include the vacuum oxygen decarburization (VOD), wherein oxygen gas is injected into the metal bath in a ladle and placed in a vacuum chamber. The argon oxygen decarburization (AOD) device, where oxygen gas and argon are simultaneously injected into the metal bath in a converter type furnace, is another type of equipment. Then, the charge in the EAF is melted down and the semi-product is manufactured, which after tapping into a ladle, followed by carbon oxidation and refining on the VOD, or after pouring from the ladle into an AOD converter, is subjected to carbon oxidation operations, refining and obtaining the finished steel.

Regardless of the type of steelmaking practices employed, the production process in an EAF includes the following periods (stages):

- furnace fettling,
- composition, preparation, and loading of the charge,
- melting the charge down (often using oxygen gas for oxidation),
- tapping.

These process periods are present in each case and for each type of practice. They comprise the whole so-called furnace melting process and refer to, in particular, the first and third group of practices mentioned above. In addition, in some cases, and especially for the second group of practices described hereinabove, there are two additional periods: oxidation and refining of the metal bath. Note that in each steelmaking shop where steel is produced on the basis of EAFs, individual engineering solutions are applied, subject to local conditions. However, the vast majority of practices can be included in the groups classified according to the above criteria. There are also steelmaking shops that use extraordinary solutions that do not directly meet the above-mentioned criteria. The types of charge materials applied in the production process and the engineering details of individual periods are discussed below.

9.1 FETTLING

Immediately after tapping the metal bath from the previous heat, the first stage of the manufacturing process of the next heat produced in the arc furnace begins, the so-called fettling. The purpose of this period is to prepare all the furnace equipment for safe and cost-effective heat production. The repair includes the assessment of the technical condition of all the structural components of the furnace, in particular the condition of the refractory lining of the bottom, walls, roof and taphole, graphite electrodes, as well as the repair of any damage. Practically after each heat, it is necessary to service the refractory lining of the walls at the slag level, and in the arc operation zone, as well as to prepare the taphole.

The damage to the monolithic part of the bottom may result from indentations from the impact of heavy pieces of scrap when loading the charge, in the case of improper preparation of the protective layer of light scrap at the bottom of the basket. The monolithic bottom layer can also be excessively worn due to overheating in the arc zone if the depth of the liquid metal and slag is insufficient in the initial melting phase, which may be the case if scrap with an inadequate density is applied. The walls of the furnace wear as a result of the erosive impact of the slag or a high intensity of the radiation of electric arcs in the case of an inadequate cover with scrap.

The refractory lining is serviced after each tapping to increase its life. Immediately after the tapping is finished, metal and slag remnants are removed from the damaged areas of the bottom, which is a condition of a good repair. Then, the actual repair is done, with the use of finely ground, calcined magnesite or dolomite. For small capacity furnaces, these materials can be thrust into the damaged areas by hand, using shovels. In medium- and large-capacity furnaces, repairs are performed with a centrifugal spreader or a belt spreader machine. A centrifugal spreader, after removing the furnace roof with an overhead crane, is put inside the furnace vessel. And then the repairing material is spread over the bottom and slopes of the vessel as a result of rotation of a disc located at the bottom of the machine. A belt spreader machine is placed outside the furnace and the repairing material is thrown in through the main door. The refractory lining of the bottom, in the area of the taphole, needs special care, as it is washed away with a stream of the flowing metal. The furnace refractory lining should be repaired as soon as possible, making sure that its cooling is as little as possible because it is only then that the good bonding of the repair material with the lining material can be obtained. Depending on the furnace capacity, the fettling scope and method of execution, the fettling usually takes from 5 to 20 min.

Regardless of the regenerative fettling, in order to increase the life of the refractory lining of the walls, protective spraying with refractory masses is applied, i.e. the so-called shotcreting. The shotcreting is carried out using materials with fine grain sizes, usually of 0.1 mm. The masses contain components increasing the adhesion of solids to water, which facilitates the formation of homogeneous water suspension, and components accelerating the sintering process. In magnesite and chromite-magnesite masses, this role is played by the previously calcined concentrate of chromium ore. The refractory mass, after spraying with water, is gunned onto the surface of the bottom, using a device called "shotcreter", wherein the necessary kinetic energy of the mass is provided by means of compressed air, with a pressure of at least 4.0 MPa, for wall shotcreting, and 6.0 MPa, for roof shotcreting.

The spraying gun is equipped with a steel lance for spraying the mass. The lance should be set at a distance of approximately 0.5 m from the surface of the bottom or walls. At a shorter distance, the mass springs back from the lining, and at a longer distance, it does not bond with the material. A single layer should be approximately 2 mm thick so that the mass does not excessively cool the refractory lining down, which would diminish the sintering effect. After the first layer has been sintered, which is assessed on the basis of the strong reheating of the brickwork, the shotcreting procedure can be repeated at the same spots. The total thickness of the applied layers during a single shotcreting operation is a few millimeters. The shotcreting is performed once per day during the whole lining campaign, or after the first symptoms of refractory lining damage occur, or in the final phase of the campaign only. In business terms, shotcreting is a very cost-effective procedure. Figure 9.1 shows a diagram of the spraying gun for the furnace walls and roof. The device consists of a vertical steel stand, attached to the furnace service platform structure. An arm made of steel structure is attached to the stand, and it can move alongside the stand. At the end of the arm, there is a moving head with a vertical boom lowered down, at the end of which there is another moving head ended with a steel tiltable pipe for spraying the shotcrete material. Thanks to a system like this, the steel pipe can move vertically and horizontally as well as rotate, which allows the sprayed material

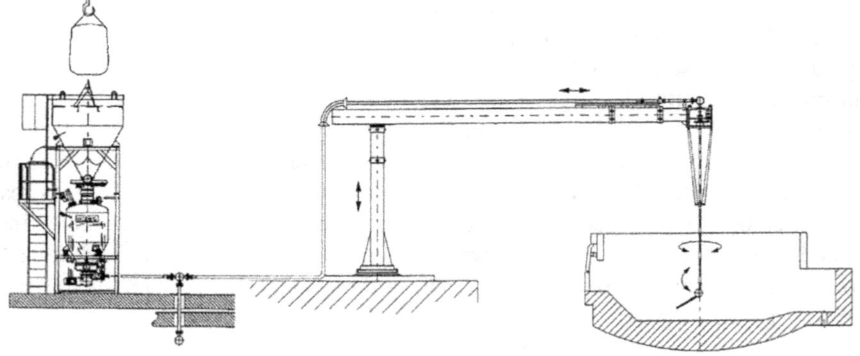

FIGURE 9.1 Diagram of a shotcreting machine for gunning the walls and roof of the furnace, along with a container for materials. (Author's work.)

stream to be directed to any place in the vessel space. The gunning material is prepared in a container station, consisting of a steel container with the material, having a cylindrical-conical shape, and a material-water mixer. The prepared material is fed with compressed air through hoses and steel pipes to the moving pipe at the end of the head. During shotcreting, the head with the pipe is placed inside the vessel, after opening the roof [2].

9.2 CHARGE COMPOSITION PRINCIPLES

Various steel grades are manufactured in the EAF, ranging from common carbon steels to high-alloy special steels, such as stainless and acid-resistant steels. In most cases, the arc furnace is used for the production of carbon steels; it is only in rare cases that alloyed steels are produced.

In the production of common carbon steels, the charge used consists of unalloyed scrap grades, and the total amount of alloying additives is small. In this case, usually unalloyed scrap is used, and the production is carried out in large furnaces with a high-power transformer installed. The charge composition issue in this case, due to the chemical composition of materials, is not very relevant. However, the selection of the scrap form and its proper arrangement in the charging basket are crucial. This is because it is decisive to reduce the possibility of mechanical damage to the bottom refractory lining during charging, and to conduct the meltdown process in an advantageous manner from the electrical and thermal perspectives.

The optimal chemical composition of the charge materials is very important for the production of high-alloy steels, due to the possibility of the recovery of alloying elements contained in the scrap that are often very expensive. The recovery technology is then used. High-alloy steels can be produced with unalloyed scrap and large quantities of suitable alloying additives in the form of ferro-alloys or technically pure metals, or by using as much alloying scrap as possible with a lower share of ferroalloys. The costs of obtaining elements contained in the scrap are much lower than the costs of obtaining elements contained in ferroalloys. Therefore, during the production of alloyed steels, it is necessary to strive for as large share of alloyed steel scrap as possible. This allows a high degree of element recovery to be obtained, and at the same time significantly reduces the costs of charge materials, which considerably reduces the total manufacturing cost of such steels [3].

The metal charge for the heat produced with oxygen gas oxidizing consists of the scrap of carbon and low-alloy steels and carburizing materials. Pig iron or the so-called carburizing agent (coke, coal, scrap of graphite electrodes, etc.) can be used as the carburizing material. The capability of carbon assimilation from pig iron by the metal bath is 100%, while from carburizing agents it is 50%–60%. Not only can pig iron be applied in the EAF charge to obtain an adequate carbon content in the metal bath, but also to reduce the content of copper, molybdenum, tin, and other nonoxidizing components in the steel produced. The proportion of carburizing materials in the charge is selected so as to obtain a sufficiently intense "boiling" of the metal bath already during the melting period. This means that the metal bath should contain, after complete scrap melting down, from 0.2% to 0.4% more carbon than the lower limit of its content in the steel grade made. The amount of carbon oxidized during the scrap meltdown period depends on the range of processes of the removal of undesired elements, such as phosphorus or sulfur, carried out at that time. However, an adequate carbon content in the metal bath after the meltdown is necessary for a correct metal bath refining process performance.

The silicon content in the charge is not limited. However, one should remember that, during the meltdown period, silicon practically and completely oxidizes to SiO_2, a chemical compound that passes into the slag. If the silicon content in the scrap is too high, a large amount of silica will be formed, which will affect the properties of the slag. Therefore, it is recommended that the silicon content in the charge does not exceed 0.5%. Proportionally to the silicon content and the potential share of pig iron in the charge, in order to obtain slag with a basicity ranging from 1.7 to 2.2 during the meltdown period, it is necessary to put into the furnace, together with the metal charge, a correspondingly greater amount of lime.

During the meltdown period, 50%–70% of the manganese contained in the charge is oxidized. However, the oxidation of manganese is a loss of this valuable element, and at the same time, it causes the formation of manganese oxides passing into the slag, increasing its mass unnecessarily. Thereby, the preferable manganese content in the charge should not exceed 0.4%. The allowed manganese content in the charge is also related to its chromium content. This component, like manganese, oxidizes in the amount of approximately 40%–60% during the charge meltdown. The resulting chromium oxides passing into the slag increase its viscosity, which deteriorates the general physicochemical properties of the slag. Chromium is also a valuable element and its oxidation is uneconomical. At the same time, at the total content of manganese and chromium exceeding 0.5%, the correct process of carbon oxidation is disturbed, which also affects the course of the steel production procedure. For this reason, the total content of these components in the charge should not exceed 0.6%.

The content of phosphorus and sulfur in the charge should be as low as possible. Virtually all grades of manufactured steels contain significant restrictions concerning the acceptable contents of these elements. The reason is the influence of these elements on the deterioration of the mechanical properties of steel. Usually, the content of these elements in the charge do not exceed 0.03%–0.06%. Bear in mind that the carburizing materials in the charge significantly contribute to an increase in the content of both of these elements in the melt. The pig iron used for the charge may contain up to 0.25% of phosphorus, while carburizing agents up to 0.8% of sulfur. Both phosphorus and sulfur can be oxidized relatively easy in the steelmaking process, but this operation requires additional amounts of oxygen, increases the amount of slag, and sometimes requires slag replacement. Consequently, it extends the steelmaking process time and causes an increase in the production costs.

The components not oxidizing in the steelmaking process may be present in the charge in amounts not exceeding their acceptable contents in the steel grade made. Elements like this in the steelmaking process include copper, nickel, molybdenum, and cobalt. If steel grades produced contain small amounts of these elements, scrap containing them can be used as the charge. However, bear in mind that their contents in the charge must not exceed their average content in the steel grade made. If the contents of these elements in the scrap are lower, they can be made up with additives of appropriate ferroalloys or technical metals to the charge.

The other alloy elements of the steel, which oxidize to a lesser or higher extent, particularly tungsten, vanadium, and titanium, should not be added with the charge. These elements, as manganese and chromium, are valuable (expensive), and they oxidize during the steelmaking process, causing the deterioration of the physicochemical properties of the slag. Therefore, such components should be introduced into the metal bath during the steel refining period, at the final stage of the melting in the furnace, or outside the furnace – in ladle metallurgy processes. The scrap containing these elements should be used in the recovery methods of steel production in order to recover them in the steelmaking process.

The charge, apart from the metallic and carburizing materials discussed above, must also contain slag-forming materials. The main slag-forming component is metallurgical lime, so-called quicklime, added in the amount of 1%–3% (10 kg/t–30 kg/t) in relation to the weight of the metal charge. Sometimes, dolomite lime is added to increase the MgO in the slag to protect the refractory lining of the walls.

When alloy steels are made in the EAF, recovery methods are usually used, without or with limited oxidizing with oxygen gas. A characteristic feature of practices like this is remelting scrap of alloyed steels in order to maximize the recovery of expensive alloying elements. Then, the charge should consist of 70%–100% scrap of the steel grade made. It should be made up to 100% with low-carbon steel scrap and appropriate ferroalloys. In both cases, 1%–3% quicklime is added to the charge, in relation to the weight of the metal charge.

The principles of charge composition are then as follows. The carbon content in the charge should be similar to its content in steel grade made. Too high a carbon content in the metal bath after the meltdown forces oxidation and undesired oxidation of the metal bath constituents. In such cases, low-carbon steel scrap is added to the charge. If the carbon content in the alloying scrap is insufficient, it can be made up with graphite electrode scrap.

The manganese content in the charge should not exceed its upper allowed limit in the steel grade made. Despite some oxidation of manganese during the scrap meltdown, this element is reduced back from the slag during the refining period. This is particularly valid in the production of alloy steels with a low manganese content. The silicon content in the charge should be as low as possible, particularly when carrying out limited refining. This delays the carbon oxidation process and reduces the basicity of the slag.

The recovery methods of steelmaking processes are characterized by adverse conditions for the dephosphorization of the metal bath. Therefore, the content of phosphorus in the charge must not exceed the upper acceptable limit in the steel grade made. When composing the charge, the phosphorus contained in ferroalloys, which may be added sometimes at the end of the melting, should also be taken into account. The sulfur content in the scrap of alloy and low-carbon steels usually does not exceed 0.03%. Since it is possible to remove sulfur from the metal bath during the refining period, no special restrictions are imposed on its content in the charge.

The content of chromium in the charge should not exceed the upper limit of the steel grade made. The content of nickel, cobalt, tungsten, molybdenum, and copper should be set at their average content in the steel grade made, making up any shortages with ferroalloys added to the charge before melting. Shortages of elements, such as chromium, manganese, silicon, vanadium, and titanium, are made up using appropriate ferroalloys when refining the metal bath under deoxidized slag or immediately before tapping.

The cost of metal materials included in the scrap is an important aspect, apart from the chemical conditions for the prepared shares of individual types of these materials. In the industrial conditions, the optimization of charge selection with regard to cost, taking into account its quality, plays a fundamental role in the profitability of production. Therefore, various calculation methods are used to compose the appropriate charge (these methods are often called charge burdening). The purpose of burdening is to select such amounts of individual charge materials to obtain the preset contents of subsequent elements keeping the cost as low as possible. Mathematically, this task can be written as shown below. A system of inequalities can be written for each element present in the steel grade made [3]:

$$a_{\min} \le \sum_{i=1}^{n} x_i a_i \le a_{\max} \tag{9.1}$$

where:

$a_{\min}$ – minimum content of a given element in the charge, %,
$a_{\max}$ – maximum content of a given element in the charge, %,
a_i – content of a given element in the charge material i (scrap, ferroalloy), %,
x_i – share of charge material i in the whole charge mass,
n – quantity of charge materials.

The share of the charge material can be expressed as a fraction, with the sum of x_i equal to 1, or as a percentage (then x_i in equation [9.1] must be divided by 100 and the sum x_i must be equal to 100). The sum $x_i \cdot a_i$ from equation (9.1) stands for the amount of element i in the charge. For instance, for carbon in the four types of scrap used as the charge materials, the equation (9.1) will take the form:

$$C_{\min} \le C_1 x_1 + C_2 x_2 + C_3 x_3 + C_4 x_4 \le C_{\max} \tag{9.2}$$

where:

$C_{\min}$ – minimum carbon content in the charge, %,
$C_{\max}$ – maximum carbon content in the charge, %,
C_1–C_4 – carbon content in the charge material 1–4 respectively, %,
x_1–x_4 – share of material 1–4 in the charge.

After writing down the system of inequalities (9.1) for all the elements present in the steel grade made, we obtain a system of inequalities, in which the shares of individual charge materials x_i are unknown. This system of inequalities can be written as follows:

$$C_1 x_1 + C_2 x_2 + C_3 x_3 + C_4 x_4 \ge C_{\min}$$

$$C_1 x_1 + C_2 x_2 + C_3 x_3 + C_4 x_4 \le C_{\max}$$

$$\mathrm{Mn}_1 x_1 + \mathrm{Mn}_2 x_2 + \mathrm{Mn}_3 x_3 + \mathrm{Mn}_4 x_4 \geq \mathrm{Mn}_{\min} \qquad (9.3)$$

$$\mathrm{Mn}_1 x_1 + \mathrm{Mn}_2 x_2 + \mathrm{Mn}_3 x_3 + \mathrm{Mn}_4 x_4 \leq \mathrm{Mn}_{\max}$$

$$\mathrm{Si}_1 x_1 + \mathrm{Si}_2 x_2 + \mathrm{Si}_3 x_3 + \mathrm{Si}_4 x_4 \geq \mathrm{Si}_{\min}$$

$$\mathrm{Si}_1 x_1 + \mathrm{Si}_2 x_2 + \mathrm{Si}_3 x_3 + \mathrm{Si}_4 x_4 \leq \mathrm{Si}_{\max}$$

And so on for the rest of the elements.

The vector x_i is the solution of the system of inequalities (9.3). For this to be the optimal solution, the minimum cost condition is yet to be met:

$$\sum_{i=1}^{n} x_i c_i \leq \min \qquad (9.4)$$

where:

c_i – price of the charge material i.

The problem presented above can be solved with many mathematical methods, e.g. using linear programming. The Simplex method is among the best known methods. In this method, the minimum of the cost function is sought through a series of reductions of variables. There are also modified variations of the Simplex method, which can be applied in some specific cases of charge burdening in EAFs. It can be, for instance, charge composing for the production of common carbon steels based on a small number of scrap classes, usually commercial scrap. In the industrial practice, various companies also use different mathematical methods to optimally burden the charge [4].

The recovery practice, due to the lack of an oxidation period, prevents hydrogen from being removed from the metal bath. Therefore, care should be taken that the physical condition of the added charge materials is appropriate (e.g. nonoxidized scrap, slag-forming materials without moisture). The rules of composing the scrap in terms of its bulk density and its arrangement in the furnace vessel are similar to those for melting with the oxidation period. One should only bear in mind that the hard-melting materials should be near the electrodes, and that the components that evaporate easily should be away from the electrodes, near the walls and slopes.

Apart from the chemical composition, the bulk density of the scrap and its arrangement in the furnace are of great importance for the proper course of the charge meltdown process [4]. According to the operating experience, the most favorable conditions are provided by a charge composed of 20%–30% of light scrap, arranged in the basket so that it lands on the bottom of the furnace vessel, thereby protecting the bottom against mechanical damage from heavy and medium scrap falling during charging. In addition, the charge should contain about 40% of heavy scrap, placed immediately under the electrodes. The balance, or 30%–40%, should comprise medium scrap, placed next to the furnace walls.

The quality of the charge materials applied has a significant impact on the performance of the steelmaking shop, and the content of trace elements in the steel produced. In this respect, the so-called primary scrap has the highest quality and usefulness. It is a process waste generated during steel manufacturing, usually characterized by a known chemical composition, and a low content of trace elements and nonmetallic impurities. Commercial, and often post-consumer scrap from metal processing plants, can contain a lot of random nonmetallic impurities and trace elements. In addition, it is often contaminated with greases and oils, and it usually features a low bulk density. To avoid unforeseen, undesirable charge components, scrap chemical analysis is often performed prior to loading, using portable analyzers based on spectral methods.

9.3 CHARGE LOADING

The charge, composed by taking into account the above criteria, is loaded into the furnace using special self-discharging baskets. The baskets designed for loading small- and medium-capacity furnaces are usually closed at the bottom with a dome made of articulated segments (Figure 9.2). These segments are made of steel plates mounted on pins, attached to the cylindrical part of the basket. At the bottom, they are ended with steel eyelets through which the fastening rope is threaded. The rope ends with a snap lock, which is opened by pulling the chain attached to the release

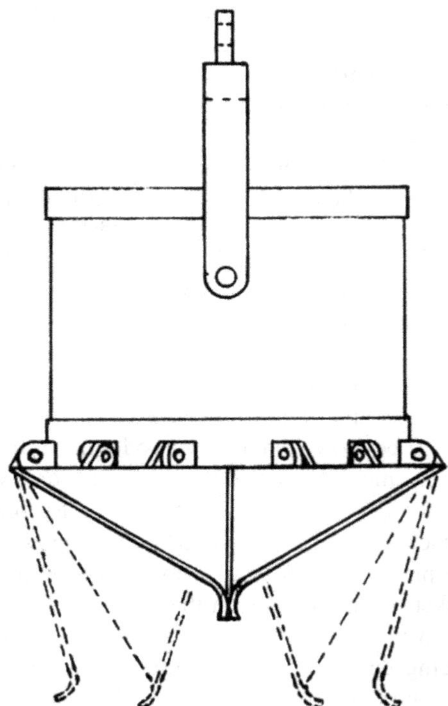

FIGURE 9.2 Diagram of the basket for charge loading with articulated (orange peel) segments [5]. (Author's work based on [5].)

spring. In the top cylindrical part, the basket is equipped with a stiffening ring, to which lugs are attached for its transport by an overhead crane. In the bottom cylindrical part of the basket, there is a stiffening ring that is also used for its placement in the charging car. The charging car has a dome-shaped structure made of steel sheets, the shape of which corresponds to the shape of the basket closed at the bottom. This facilitates fastening flexible segments with a rope because, when it is placed on the car, they automatically converge inward.

Medium or large capacity furnaces are loaded with baskets closed at the bottom with semi-dome segments tilted sideways (Figure 9.3). Two semi-dome segments are made of steel sheets and are mounted on bolts in the bottom cylindrical part, which allows them to be tilted sideways in order to discharge the scrap. Tilting is performed by means of steel cables attached to the crane hook. A basket like this is also placed in the charging car, while scrap is being loaded into it in the charging bay. After the scrap has been loaded, the basket is transported on a charging car (sometimes using an overhead crane) to the vicinity of EAF. An example of the view of the clamshell-type loading basket, located on the charging car, is shown in Figure 9.4.

Scrap is loaded into the basket from the bins with grippers suspended on cranes. Usually, electromagnetic grippers are used, to which scrap metal is "hooked" under the influence of the magnetic field forces generated around the gripper as a result of current flowing through coils placed inside it. An example of such a gripper is shown in Figure 9.5.

The method of arrangement of the charge in the basket should correspond to its future arrangement in the vessel because during the loading operation there is no significant change in the nature of this arrangement. The scrap arrangement should enable the liquid metal bath covered with slag to be rapidly formed, and at the same time prevent the mechanical damage to the refractory lining of the furnace bottom

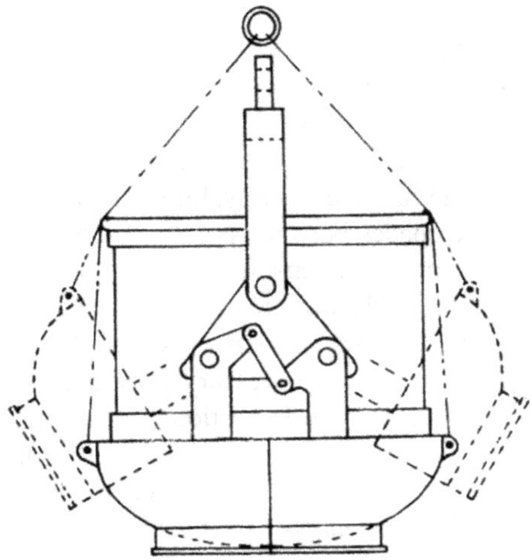

FIGURE 9.3 Diagram of a clamshell basket for loading charge, with semi-dome segments tilted sideways [5]. (Author's work based on [5].)

FIGURE 9.4 View of a charging basket on the charging car. (Author's photo.)

FIGURE 9.5 View of an electromagnetic gripper with suspended scrap. (Author's photo.)

during scrap loading. For this purpose, lightweight scrap metal is loaded to the bottom of the basket, followed by pig iron or other carburizing agents. These materials are covered with lime, and then heavy and high-melting scrap is charged into the arcing zone. Medium scrap and scrap containing easily evaporating components are loaded at the periphery of the furnace. Finally, lightweight scrap is charged to the top again, which should fill the spaces between pieces of thick and medium scrap. This arrangement of the charge materials ensures their good compactness, and a fairly stable operation of electric arcs during meltdown. For a single loading of the total quantity of the charge per heat, its bulk density should be at least $2.0 < t/m^3$. With a lower bulk density of the scrap it is necessary to add the charge after the partial meltdown of its first and possibly the next portion.

Recently, attempts have been made to automatically control the procedure of loading the charge into the basket. Such systems collect the information about the size of individual pieces of scrap suspended to the gripper based on the processing of images from digital CCTV cameras recording the loading process. The obtained information enables the development of data sets and classification models for controlling the procedures of placing scrap metal in the basket [6,7].

Loading of the charge with the basket into the furnace is performed in the following way. First, the roof is removed from above the furnace vessel. It is first lifted and then tilted. Using a crane, the basket is moved over the vessel and then lowered to a height of about 0.5 m above the top edge of the shell. The basket is opened and the materials, under the influence of their own weight, fall onto the bottom. After the charge is discharged from the basket, the basket is transported from above the furnace, and immediately afterward the roof is covered. When the charge is loaded in a few portions, the total amount of carburizing and slag-forming materials are added into the furnace with the first portion of the scrap or divided between the first and the second charge basket.

9.4 MELTDOWN PERIOD

9.4.1 Energy-Related Aspects

Immediately after loading the charge, and covering the vessel with the roof, the supply voltage is switched on, and the automatic control system of the electrodes positioning is activated. The lowering of the electrodes begins, and when they come into contact with the charge, the electric arc is struck, and the scrap meltdown process begins. Electric arcs burn at the tips of the electrodes, between their lower flat surface, known as the "face", and pieces of the loaded scrap. Initially, the action of electric arcs covers the charge under the electrodes, located in the upper part of the vessel – near the roof. The so-called craters are then melted in the scrap, the diameter of which is approximately 50% larger than that of the electrodes. During this time, the electric arcs are very unstable, due to an insufficient reactance of the cold scrap, and rapid changes in the resistance of the arcs as a result of a low content of the charge and rapid melt of its fine lumps, or the displacement of large lumps, which can result in short circuits. The occurring high fluctuations in the value of the voltage and current of the electric arc, even with an efficiently working automatic electrode positioning control system, may lead to breaks or short circuits, which adversely affects the supply network. In order to reduce the negative effects of short circuits and to protect the roof, the power of arcs is limited in the initial meltdown phase to 60%–80% of the transformer maximum power output. For small-capacity furnaces equipped with a choke, it is connected to the power supply system to increase its inductance.

Due to the presence of fine scrap in the upper part of the vessel, the initial meltdown period is characterized by the rapid melting of craters, "sinking" the electrodes into the scrap and the downward movement of the burning arc. After the assumed time (e.g. 2 min), or after the assumed amount of energy has been introduced into the furnace (e.g. 1,500 kWh), the maximum power of the supply transformer is switched on. The power is changed by switching to another tap of the furnace transformer with a higher voltage. Then, the electric arcs become longer, and the arc powers become higher.

After the first 2–3 min, the craters are melted in the scrap to a depth of 0.6–0.8 m from the roof, the scrap is quickly heated to higher temperatures, and the conditions of arcing become more stable. In addition to melting downward, thermal energy is radiated from the arcs onto the surrounding charge. The simultaneous intensive melting of scrap in width begins. The burning electric arcs are surrounded by a solid metallic charge and, therefore, there is no possibility of the direct radiation of thermal energy to the vessel walls. Therefore, there is no concern of overheating the refractory lining or damage to the water-cooling elements of the furnace walls. Intensively radiating electric arcs quickly heat up the charge, which at the same time caves in under the influence of its own weight. From the beginning of the arc burning, the molten metal flows down toward the hearth, and the formation of the metal bath begins. The level of the metal bath increases as the scrap melts. Throughout the meltdown process, an automatic electrode positioning system controls the position of the electrodes, to maintain the desired length of the burning arc. Initially, the arcs burn between the electrode tips and pieces of scrap. As the amount of melting scrap decreases, sometimes larger pieces sink and they rapidly move. Then, short circuits or breaks of the arc can appear, which causes breaks in its burning. At this stage, the arcs continue to burn in an unstable manner. This melting period, depending on the power of the supply transformer, lasts 8–12 min.

After melting the craters and forming a liquid metal bath, the arcs burn between the electrode tips and the liquid metal. Then, the arc burning conditions definitely stabilize, and as the level of the metal bath rises, the electrodes move upward relatively gently. As of this moment, the melting of the charge takes place only in width.

The maximum power of the arcs is maintained until a certain amount of energy is introduced into the furnace, e.g. 250 kWh/1 ton of charge. After that, the power of the supply system is reduced to a level depending on the scope of the slag foaming practice. This is related to the protection of the furnace walls against the impact of the electric arc. If the slag foaming practice is not used, the power is reduced, for instance, to about 60% of the maximum, and in the case of "full" foaming of the slag, the power can still be close to maximum. The arcing with the reduced power is maintained until the end of the conventional meltdown period of the scrap from the basket. The end of the conventional meltdown period means the moment when most of the scrap is melted, the arcs are already exposed, and there are still single solid pieces of scrap near the walls above the metal bath. Then, the power supply system is isolated, the electrodes are lifted, the roof is lifted and tilted, and the loading of the second basket can be started. The procedure for melting the second, and potentially subsequent baskets, is the same as for the first one.

The total time of the charge meltdown period, depending on the power of the supply transformer, is 30–100 min. Figures 9.6 and 9.7 show a graphic view of the various stages of the charge meltdown in the arc furnace. In the process of scrap meltdown in the EAF, the following phases of melting each portion of the charge from successive baskets can be distinguished:

- The initial phase occurs from the beginning of meltdown (switching on the electric power supply) until the melting of the craters in the scrap.
- The second phase covers the meltdown of most of the scrap and lasts from the time of melting the craters until reaching the level of the metal bath, including melting the scrap in width, until the burning arc is exposed (then, the maximum arc power is used).

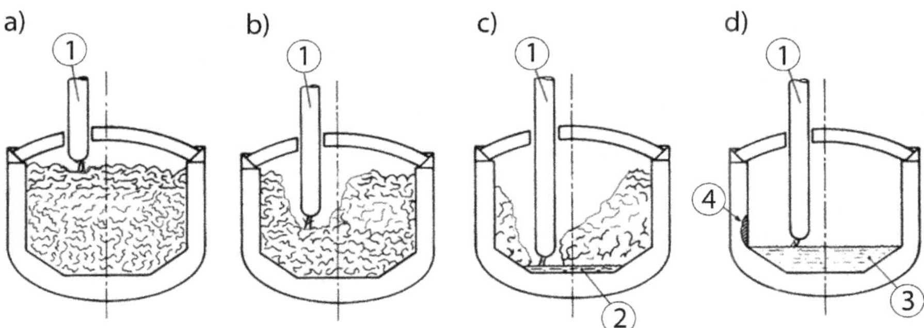

FIGURE 9.6 Diagrams of individual stages of charge meltdown in the arc furnace: (a) arc initiation, (b) crater melting, (c) arc reaching the level of metal bath, (d) heating the bath after scrap melting: 1 – electrode, 2 – initial metal bath, 3 – metal bath, 4 – hot spot. (Author's work based on [5].)

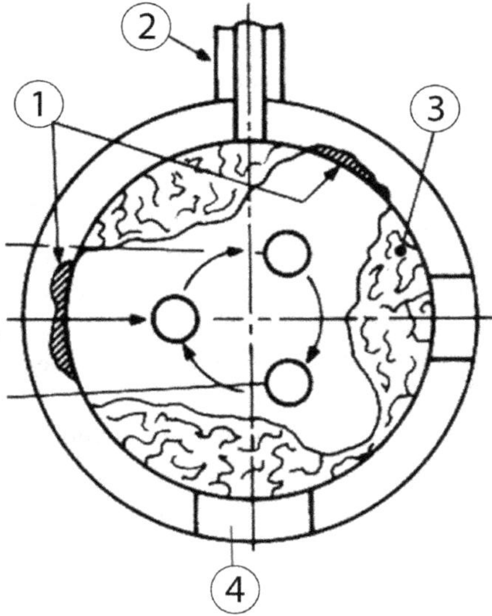

FIGURE 9.7 Diagram of the top view, charge meltdown in the furnace at the stage of melting in width: 1 – hot spot, 2 – pouring spout, 3 – unmelted (solid) pieces of scrap, 4 – main door. (Author's work based on [5].)

- The third phase includes melting the scrap metal located near the peripheries of the vessel (at the end of this phase, another charging basket is loaded).
- The final phase is melting the remaining solid scrap pieces near the vessel walls, and the so-called reheating of the metal bath to the proper temperature.

Taking into account the above-described conditions related to the charge meltdown technology, in specific cases of the arc furnace operation, the rules for determining the electrical parameters (so-called operating states) of the power supply system are developed. These principles are called power programs. They are aimed at optimizing technical, thermal, and business conditions of scrap meltdown. The selection of power programs in the industrial conditions is carried out by experimental verification of the following rules:

1. The actual amount of energy needed to melt scrap from each basket (or per 1 metric ton of charge) down is divided into three parts:
   - the amount needed to "start-melting" the charge and pre-melting the craters (about 5%–10% of the energy needed to melt the scrap from a given basket down),
   - the amount needed to melt the main part of the charge (one portion in the amount of 75%–80% or two portions in the amount of 35%–40% of energy needed to melt the scrap from a given basket down),
   - the amount needed to finally melt the remaining part of the charge down, and to heat the metal bath to the proper temperature (about 10%–15% of the energy needed to melt the scrap metal from a given basket down).
2. Voltages and the numbers of the taps of the furnace transformer on which to work at each stage of the charge meltdown are determined.
3. The times of tap change-overs, according to the electricity consumption or as a function of time, are determined.
4. The so-called chemical energy from the fuel combustion in the burners, and the oxidation of the components of the metal bath must be taken into account.
5. Consideration should be given to increasing the energy efficiency of heat transfer of the arcs when using the slag foaming practice.

An example of a power program diagram for the charge meltdown period in the EAF is shown in Figure 9.8.

In order to shorten the scrap meltdown period, methods intensifying this process are applied from the moment of switching on the maximum power of the furnace transformer. The most common methods are oxygen gas injection with lances and the use of gas-oxygen burners.

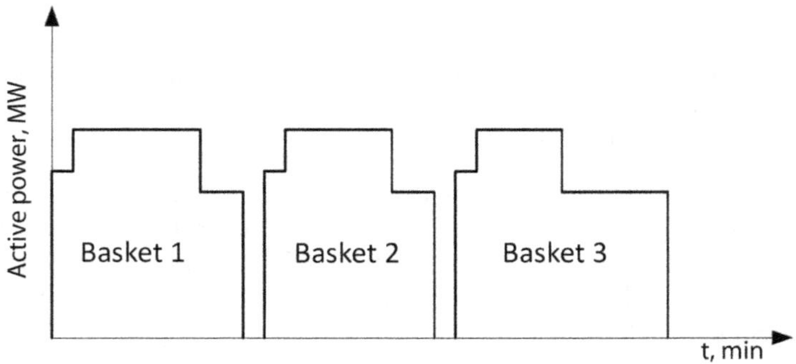

FIGURE 9.8 Example of a power program diagram for the charge meltdown period in the EAF thermal energy transfer efficiency. (Author's work.)

The lances are placed in the main door and at the initial stage of melting the scrap from each basket down (when the vessel is completely filled with it), they are set so that the oxygen jet is directed toward the scrap near the main door. Although the oxygen jet injected onto the surface of the solid scrap primarily oxidizes iron, it is a strongly exothermic reaction, contributing to the rapid heating and melting the scrap within this area. Despite some iron losses, the associated reduction of the tap-to-tap time reduces heat losses, which brings about a measurable economic effect in the steel production.

As the metal bath forms, and the scrap meltdown is sufficiently advanced, which allows the introduction of the lance into the melt, the oxygen injection into the volume of the liquid metal begins. Then, the components of the metal bath are oxidized, including the most important element of the liquid ferrous solution – carbon. Carbon is the only element in the liquid ferrous solution (metal bath) which, as a result of oxidation reaction, generates a gaseous product (carbon monoxide) within the metal bath volume. As a result, the CO bubbles forming in the bath volume, flowing upward, stir the bath. This phenomenon is very advantageous for homogenizing the temperature within the bath volume, accelerating the process of melting the scrap lying in its cold zones down. With the high kinetics of the reaction of carbon oxidation with oxygen in the metal bath, the process of CO gas bubbles production is so intense that this phenomenon resembles boiling.

The oxygen consumption for the intensification of charge meltdown and the subsequent oxidation depends on the type of scrap, the expected scope of oxidation of carbon and other components of the metal bath, the type of practice applied, and the steel grade produced. In the industrial practice, the consumption of oxygen gas ranges from 15 to 40 Nm^3, per 1 metric ton of steel produced. The flow rate of the oxygen injected is between 15 and 60 Nm^3/min., depending on the furnace size and the carbon oxidation extent. At the initial phase of the scrap meltdown period, lower oxygen flow rates are applied, while after complete meltdown, during oxidation, higher flow rates are applied. Lances, with parameters customized for each furnace, usually have adjustable oxygen flow rates within a certain range, to be selected according to the process needs during.

The operation of gas-oxygen burners, particularly during the scrap meltdown, is related to the efficiency of transferring the thermal energy of the burner flame to the scrap. This efficiency is highest at the beginning of the meltdown period when the scrap has a low temperature and a substantial surface area. As the meltdown time progresses and the scrap temperature increases, the burner performance decreases. For this reason, the burners are activated from the beginning of the charge meltdown period. They operate at the maximum power for a period of 60%–70% of the meltdown time of the first basket, and 30%–50% of the meltdown time of the subsequent baskets, depending on the type and form of the scrap. After that, their power is turned down to a level depending on the form of scrap. After melting the charge down, the burners usually operate at their minimum power, approximately 5%–15% of the maximum power. The demand for natural gas, depending on the power of the burners, is 5–20 Nm^3 of gas per 1 metric ton of the steel produced. The demand for oxygen depends on the adopted ratio of excess oxygen; it amounts to 10–30 Nm^3 of oxygen per 1 metric ton of the steel produced.

In a sense, the post combustion of carbon monoxide in the upper part of the vessel, especially in the area of the exhaust gas outlet, can be incorporated into the system that intensifies the charge meltdown in the arc furnace. This method is used at the end of the meltdown period when most of the scrap is molten and the upper part of the vessel is "exposed". Post combustion of CO is carried out by injecting oxygen through a lance directing its stream into the upper regions of the vessel. The oxygen flow rate used is between 5 and 10 Nm^3/min, depending on the size of the furnace and the extent of the carbon oxidation. The intensifying methods also include the technology of blowing inert gases through porous plugs installed in the bottom of the furnace. This method is not intended to provide any additional chemical energy to the system. It only performs the function of stirring the bath by providing kinetic energy, which accelerates the melting of the scrap contained in the bath volume.

Although in a modern arc furnace, the thermal energy needed for the metallurgical process is obtained mainly from the electricity, the chemical energy obtained from the combustion of natural gas, from the oxidation of iron from scrap, from the oxidation of metal bath components (mainly silicon and carbon), and post combustion of carbon monoxide also plays an important role. The entire EAF energy system consists of many pieces of equipment, including the power supply system, gas-oxygen burners, oxygen lances, and lances for coal and lime injection. The operation of each of these devices is numerically controlled to maximize the utilization of the energy obtained.

The electric power control is divided into two systems:

- electric arc power control, which is executed by changing the taps in the furnace transformer, and changing the current intensity within each tap,
- maintaining the assumed arc length, which is achieved by changing the position of the electrode support arms.

As regards controlling the operation of burners and lances, there are also two systems:

- gas and oxygen flow rate control, respectively,
- control of the moment of switching on and the time of changes in the flow rate of gas and oxygen, respectively,
- control of the position of the lances.

The operation of burners and lances is usually controlled on the basis of the measurements of temperature and chemical composition of exhaust gases in the furnace elbow.

For the purpose of optimal selection and cost-efficient utilization of individual types of energy to melt the charge down, and subsequently conduct the metallurgical process of steel production, comprehensive numerical control systems are used to control the operation of all the facilities supplying energy utilities [7–9]. In an industrial practice, there are furnaces not equipped with gas-oxygen burners, wherein only one oxygen lance is used to oxidize the metal bath. In this case, the energy demand for the process comes mainly from electricity, and the share of chemical energy is a few percent. Modern furnaces are usually equipped with several burners and two or three lances for injecting oxygen, and lances for injecting coal and lime. The demand for thermal energy in such a case comes only in 50%–70% from electricity and 30%–50% from the chemical energy. The application of methods intensifying the charge melting process contributes to the reduction of electricity consumption, shortening of melting time, and the reduction of graphite electrode consumption.

9.4.2 Oxidation of Metal Bath Components

Simultaneously with the charge meltdown, the process of forming the metal bath, dissolving the carbon introduced with the charge in the form of lumps, oxidation of the metal bath components, and the process of slag formation begins. Slag is formed from the products of oxidation reactions of additions, nonmetallic impurities of scrap (added to the furnace together with the metal charge), lime, and wearing out the refractory lining. Oxidation of additions takes place with oxygen injected through the lance and partially with oxygen from the air in the atmosphere inside the furnace. On the surface of the heated scrap, from the beginning of the meltdown period, the iron contained in the scrap oxidizes according to the reactions:

$$2\ \mathrm{Fe} + \mathrm{O_2} = 2\ \mathrm{FeO} \tag{9.5}$$

$$3\ \mathrm{Fe} + 2\ \mathrm{O_2} = \mathrm{Fe_3O_4} \tag{9.6}$$

Iron oxide is often further oxidized at the process temperature according to the reaction:

$$6\ \mathrm{FeO} + \mathrm{O_2} = 2\ \mathrm{Fe_3O_4} \tag{9.7}$$

These reactions are exothermic with a considerable thermal energy release. The oxides formed feature a high evaporation pressure at the temperatures prevailing inside the furnace. Some of them evaporate and are evacuated together with the waste gases in the form of brown smoke. The resulting iron losses do not exceed 0.5%–1.5% of the charge weight. At the same time, the rest of the oxides that have formed already at the surface of the heated scrap constitutes the beginning of the liquid slag, which takes part in the oxidation reactions of other charge components. These reactions occur at the surface of the melted scrap, and at the interface between the liquid metal bath and the slag. During this time, primarily silicon, phosphorus, manganese, and chromium, and to a much lesser extent the carbon oxidize and the desulfurization of the bath occur. These processes run according to the following reactions:

$$[\mathrm{Si}] + 2\,[\mathrm{O}] + 2\left(\mathrm{O}^{2-}\right) = \left(\mathrm{SiO}_4^{4-}\right) \tag{9.8}$$

$$2\,[\mathrm{P}] + 5\,[\mathrm{O}] + 3\left(\mathrm{O}^{2-}\right) = 2\left(\mathrm{PO}_4^{3-}\right) \tag{9.9}$$

$$[\mathrm{Mn}] + \left(\mathrm{Fe}^{2+}\right) = \left(\mathrm{Mn}^{2+}\right) + [\mathrm{Fe}] \tag{9.10}$$

$$[\mathrm{Cr}] + \left(\mathrm{Fe}^{2+}\right) = \left(\mathrm{Cr}^{2+}\right) + [\mathrm{Fe}] \tag{9.11}$$

$$[C]+[O]=\{CO\} \tag{9.12}$$

$$[S]+\left(O^{2-}\right)=\left(S^{2-}\right)+[O] \tag{9.13}$$

Equilibrium constants and kinetic characteristics of oxidation of these elements (reactions 9.8–9.13) are related to the range of iron oxidation (reactions 9.5 and 9.6), as well as the amount and the properties of the lime introduced with the charge. Oxides, which dissolve in iron oxides, are the products of the above-mentioned reactions. At the same time, the lime is dissolved in the forming oxide phase. The above processes occur particularly and intensively in the area of electric arc impact, where the prevailing temperatures are much higher. The flowing-down products of reactions (9.5)–(9.13), including the processes of mutual dissolution, form at the surface of the forming liquid metal bath, an oxidizing, basic slag, called the primary slag. The properties of such slag help stabilize the conditions of electric arc burning, which increases the rate of charge melting and heating of the metal bath. Higher temperatures at the same time increase the kinetics of the oxidation reaction of metal bath constituents.

Consequently, the processes described hereinabove cause more intensive stirring of the metal bath, occurring as a result of convective currents caused by the heterogeneity of the temperature field, and the carbon oxidation reaction. Convective stirring of the bath is of limited importance despite significant temperature differences. A much greater role in the bath stirring is attributed to the carbon oxidation reaction. When the charge is melted in the furnace, this reaction occurs on the capillaries of the refractory lining of the bottom or pieces of the melting scrap. Carbon monoxide gas bubbles (according to the reaction 9.12) nucleate at these points and begin to flow upward through the metal bath after reaching critical dimensions. When flowing out, the size of the CO bubbles increases. Their flow causes the stirring of the melt. This stirring accelerates the transport of oxygen and carbon to the place where the oxidation reaction takes place, and at the same time, it influences the course of the heterogeneous oxidation reactions of other metal bath additions at the interface of the metal bath and slag.

Due to the high affinity for oxygen, silicon is oxidized very intensively during the charge meltdown. The silicon oxidation reaction is often written down in a simplified way:

$$[Si]+2\,[O]=(SiO_2) \tag{9.14}$$

The formed SiO_2 has rather little activity in the basic slag because it is bound in stable ionic complexes with CaO. In these slags, FeO is highly active. For this reason, the equilibrium of the silicon oxidation reaction is established with very low silicon contents in the bath, which practically results in the complete oxidation of silicon in the initial period after the formation of the metal bath. The silicon oxidation reaction is highly exothermic. The silica that is formed is easily dissolved in the previously formed iron oxide, forming a compound called fayalite ($2FeO{\cdot}SiO_2$).

In the basic slag, not only does the activity of MnO and FeO depend on their contents but also on the basicity of the slag. Thereby, the oxidation conditions of the manganese during the charge meltdown are influenced by the chemical composition of the slag, and the temperature of the metal bath. In the basic slag, the activity of MnO is high, and for this reason, as the basicity of the slag increases, the equilibrium of the manganese oxidation reaction is established at higher manganese contents in the metal bath. In the same way, the manganese oxidation reaction equilibrium is influenced by a temperature increase. For these reasons, for high manganese contents in the charge, the scrap meltdown process should be carried out with strongly oxidizing slags, to achieve low manganese contents in the bath. Similarly, the temperature and chemical composition of the slag influence the oxidation of chromium. If the content of these elements in the charge does not exceed 1%, during the charge meltdown period, 40%–80% of their initial content is oxidized. Both manganese and chromium oxidation reactions are exothermic, but the amount of heat emitted in them is much less than that for iron.

Other charge components, having an affinity for oxygen higher than iron (e.g. tungsten and vanadium), are also oxidized in the meltdown process in a similar way to manganese or chromium. However, it is recommended that the content of the oxidizing components in the charge is as low as possible. On the one hand, such components are economically valuable elements, and their oxidation brings about measurable economic losses. On the other hand, as a result of oxidation, the mass of slag increases, which is technologically and economically disadvantageous (the demand for heat energy increases).

Other constituents (elements) contained in the charge, with the affinity for oxygen lower than that of iron (e.g. nickel or copper) are not oxidized in the metallurgical process. Therefore, they are transferred to the steel produced in as much as they were present in the charge. Therefore, it is important that the contents of these elements in the charge do not exceed the admissible levels in the finished steel.

The charge meltdown process is accompanied by the oxidation of the phosphorus contained in it, according to the reaction (9.9). The resulting anion $\left(PO_4^{3-}\right)$ has a strong affinity for the calcium cation Ca^{2+}, present in the forming slag. Calcium phosphates (stable at high temperatures) are formed according to one of the reactions:

$$2\,[P] + 5\,(FeO) + 3\,(CaO) = (3\,CaO \cdot P_2O_5) + 5\,[Fe] \tag{9.15}$$

$$2\,[P] + 5\,(FeO) + 4\,(CaO) = (4\,CaO \cdot P_2O_5) + 5\,[Fe] \tag{9.16}$$

The reaction (9.15) is assumed to occur at a lower availability of cations Ca^{2+} (slag basicity under 2.5), and the reaction (9.16) at a higher level (slag basicity above 2.5). At the initial meltdown stage, the reaction (9.16) is more likely to occur. The dephosphorizing of the melt occurring at this time can be characterized by the coefficient of phosphorus partition between the metal and the slag. To define this coefficient, a simplified reaction of phosphorus oxidation is assumed:

$$2\,[P] + 5\,[O] = (P_2O_5) \tag{9.17}$$

Then, the coefficient of phosphorus partition is the quotient of the amount of the phosphorus contained in the slag, bounded in the pentoxide, and the phosphorus contained in the metal:

$$L_P = \frac{(P_2O_5)}{[P]^2} \tag{9.18}$$

In the industrial conditions, a parameter being the so-called dephosphorizing degree is often used, which is defined as follows:

$$\eta_P = \frac{[P]_p - [P]_k}{[P]_p} 100\% \tag{9.19}$$

where the symbols $[P]_p$ and $[P]_k$ represent the initial and final phosphorus content in the metal bath. Taking into account the thermodynamics of the phosphorus removal reaction, it can be assumed that the dephosphorizing process is generally characterized by the following relationship:

$$L_P = \frac{(P_2O_5)}{[P]^2} = f\left((CaO),(FeO),(SiO_2),(P_2O_5),T,K_p\right) \tag{9.20}$$

where T represents the temperature, while K_P is the equilibrium constant for the dephosphorization reaction.

The melt dephosphorization is determined by FeO, CaO, and SiO_2 content in the slag (more precisely – the basicity of the slag expressed by the ratio of CaO to SiO_2). On the basis of the conducted experimental research, a correlation was developed to describe the dephosphorization process for slags containing mainly FeO, CaO, and SiO_2, with low contents of MgO, MnO, and Al_2O_3 [10]:

$$\log\frac{(P)}{[P]} = \frac{22{,}350}{T} - 16.0 + 2.5\log(Fe_c) + 0.08(CaO) \tag{9.21}$$

where (P), (Fe_c), and (CaO) stand for the phosphorus, total iron, and calcium oxide content in the slag, [P] represents the phosphorus content of the metal bath, and T represents the temperature in Kelvin degrees, respectively.

Not only is the bath dephosphorization influenced by the chemical composition of the slag, but also its quantity. The estimated content of P_2O_5 in the slag can be determined from the relationship:

$$(P_2O_5) = \frac{m_k\left([P]_w - [P]_k\right)\dfrac{m(P_2O_5)}{m_2[P]}}{m_z} \tag{9.22}$$

where:

P_w – phosphorus content in the charge, %,
P_k – phosphorus content in the metal bath, %,
m_k – mass of the metal bath, kg,
m_z – mass of the slag, kg,
$m(P_2O_5)$ – molecular weight (P_2O_5), 142 g,
$m_2[P]$ – atomic weight of two phosphorus atoms, 62 g.

From the relationship (9.22), and taking into account the equation (9.18), it is possible to determine the minimum weight of the slag that will be "capable of absorbing" the oxidized phosphorus from the bath, without knowing the content of P_2O_5 in the slag, assuming the phosphorus partition coefficient between the metal and the slag. The analysis of the performed calculations shows that in order to remove 0.02% of phosphorus from the charge, it is necessary to create slag in the amount of at least 10% of the weight of the metal charge. It is required to add a significant amount of lime to the process to form this amount of slag.

In classic technologies, lime is added in the form of lumps with the charge, in the amount of 1%–3% of the weight of the metal charge, which results in the formation of slag in the amount of about 5% in relation to the weight of the charge. Such conditions enable no more than 0.01% phosphorus to be removed from the metal bath. In order to remove more phosphorus, at the end of the meltdown period, the so-called racking off of the slag (which means pouring it through the main door), and adding further portions of lump lime, also through the main door, should be performed to create new slag.

In modern technological solutions, the first portion of lime is also added in the form of lumps with the charge. The remaining portion, necessary to achieve the desired degree of dephosphorization, is fed by injecting pulverized lime with a lance immersed in the slag. This solution facilitates the optimization of the metal bath dephosphorization process.

The oxidation of other components during the charge meltdown in the EAF occurs to a lesser degree. The charge meltdown period is considered complete after the scrap is completely melted and the metal bath and slag are formed, and at their surface, especially in the area of the vessel walls, no protruding pieces of scrap are visible any longer. At this point, the arc operation is interrupted by switching off the power supply and lifting the electrodes. The furnace operator starts measuring the temperature of the metal bath and taking a metal sample to analyze its chemical composition. The scope of further process operations during steel melting depends on the obtained results of the measurements and analyses.

In classic solutions, applied in some steelmaking shops, and particularly in foundries equipped with small arc furnaces, the further manufacturing process includes subsequent stages aimed at producing finished steel, which is characterized by an appropriate chemical composition and proper metallurgical cleanness as well as an adequate temperature allowing the casting to be performed correctly. Plants of this type are not equipped with secondary metallurgy facilities, and the technological stage following the charge meltdown is the oxidation period of the metal bath.

In modern solutions, used in steelmaking shops equipped with medium- and large arc furnaces, primarily producing common carbon steels, the downstream process includes only the tapping of the liquid metal bath obtained after melting the charge. Further process steps leading to obtaining finished steel with an appropriate chemical composition, and the assumed metallurgical cleanness, and obtaining the proper temperature for its casting, are carried out in various types of secondary metallurgy facilities, mainly in the ladle furnace. In this case, after measuring the temperature, depending on the result obtained, a decision is made as regards further heating the bath in order to obtain a value consistent with the assumed engineering criteria. If the obtained temperature measurement shows a value that meets the criteria, then from this point of view, the bath, as a semi-product, is ready to be tapped. The second criterion for the tapping readiness is the chemical composition. The engineering criteria assume the acceptable contents of individual elements in the bath. The most common restrictions relate to the content of carbon as well as alloying elements and undesirable elements. As it is not possible to oxidize the metal bath components in secondary metallurgy facilities (in general), the acceptable content of all elements must not exceed their acceptable content in the finished steel. This concerns in particular undesirable elements such as copper, tin, phosphorus, and sulfur.

In most cases, the metal bath, which after melting is tapped into the ladle for further refining in secondary metallurgy facilities, is characterized by the following chemical composition: 0.05%–0.10% carbon, up to 0.05% silicon, up to 0.10% manganese, up to 0.02% phosphorus, up to 0.02% sulfur, and the minimum content of other elements. The tapping temperature is within the limits of 1,853–1,923 K (1,580°C–1,650°C).

9.5 OXIDATION PERIOD

Currently, in industrial practice, the metal bath is oxidized with oxygen gas, using a lance in the form of a steel pipe immersed in a metal bath or a lance equipped with a supersonic nozzle (de Laval), placed above the bath. The classic pipe lance is entered into the furnace through the main door and immersed in the liquid metal into a depth of about 200 mm. The injected oxygen jet penetrates directly into the metal volume. The supersonic lance can be entered through the main door or placed in the roof. It is water cooled and is not immersed in the bath. It enables the oxygen jet velocity to be increased to a supersonic value. The oxygen jet is directed onto the bath from a certain distance, but thanks to the high kinetic energy it penetrates deep into the volume of the liquid metal, enabling good oxygen penetration, and the intensification of the ongoing oxidation reactions. The applied working pressure of oxygen in the lance is 0.6–1.2 MPa and its consumption is 6–15 Nm^3/t of the charge.

Depending on the method of supplying oxygen, its jet carries different kinetic energy. In the case of injecting the oxygen jet directly into the metal bath volume, the oxygen intensively dissolves in the liquid iron solution. In this area, the temperature is much higher (about 2,673 K (2,400°C)), which is associated with an increase in the solubility of oxygen in the solution, to the level of 2% or even more. The temperature in the other areas of the bath is much lower (about 1,873 K (1,600°C)), which is related to the about ten times lower solubility of oxygen in the solution – the level of 0.2%.

Therefore, during the movement of oxygen to the areas with a lower temperature, oxygen is released from the solution in the form of FeO with a highly developed surface, which favors the oxidation processes. In the case of supplying a supersonic jet to the surface of the bath, its high kinetic energy causes the formation of a crater, from which metal and slag droplets eject, creating an area with an emulsified metal-slag-gas phase within this zone, with a very developed contact surface. The direct contact of metal droplets with oxygen bubbles enables iron to oxidize intensively, and the FeO formed as a result of this reaction is transferred to other regions of the metal bath, thanks to turbulent pulsations of its entire volume. Under such conditions, the iron oxide becomes a kind of "oxygen source" in the bath areas that are remote from the oxygen jet [11].

The main purpose of the oxidation of the metal bath is to reduce impurities (phosphorus, sulfur, nitrogen, hydrogen, etc.) to a level that guarantees the proper quality of the steel produced, carbon oxidation to the level corresponding to the grade of steel produced, and to heat the bath to a temperature necessary for deoxidation, making up the chemical composition and heat tapping. After the charge is completely melted, the power is reduced to about 50%–70% of the rated power of the furnace transformer, metal and slag samples are taken to analyze their chemical composition. Then, about half of the melting period slag is removed from the furnace, and its chemical composition is adjusted by adding the appropriate amount of lime to the furnace. At the beginning of the oxidation period, the slag should have basicity within the range of 2.5–3.0, which is necessary to create good conditions for the dephosphorization and desulfurization of the bath. After adjusting the chemical composition of the slag, the stage of intensive oxidation of the metal bath begins.

The oxygen gas supplied directly to the volume of the metal bath causes intense oxidation of iron and other components:

$$[\mathrm{Fe}]+[\mathrm{O}]=(\mathrm{FeO}) \tag{9.23}$$

$$[\mathrm{Mn}]+[\mathrm{O}]=(\mathrm{MnO}) \tag{9.24}$$

$$2\,[\mathrm{Cr}]+3\,[\mathrm{O}]=(\mathrm{Cr_2O_3}) \tag{9.25}$$

The silicon introduced with the charge oxidized according to the reaction (9.14) already during the meltdown period. All oxidizing reactions (9.23)–(9.25) are exothermic, and their course is accompanied by a rapid increase in the metal bath temperature, which favors the intense oxidation of carbon at a rate exceeding 1% C/h, and at the same time hinders the oxidation of manganese and chromium. The oxidation of the bath with oxygen gas facilitates achieving a carbon content under 0.02%. In such conditions, the time of the oxidation period does not exceed 20 min, and the electricity consumption is not more than 15 kWh/t of steel manufactured. During the oxidation period, the slag is intensively foamed.

Often, particularly when intensive practices of low carbon, phosphorus, and sulfur steel manufacturing are applied, the injection of lime into the metal bath is carried out during the oxidation period. This intensifies the oxidation of carbon and significantly accelerates the process of the dephosphorization and desulfurization of the metal bath [12].

9.5.1 Carbon Oxidation

The oxidation of carbon and other elements, components of the metal bath, is a complex, multistage process of chemical reactions. Without going into detailed theoretical considerations, it is assumed that the oxidation of carbon in the arc furnace conditions during the oxidation period occurs with the participation of carbon and oxygen dissolved in the liquid ferrous solution, with the formation of gas bubbles of CO, and occurs according to the general formula of the reaction (9.12). Admittedly, a small part of the carbon can oxidize to CO_2, but under industrial conditions, the scope of such a reaction is very limited, and in a practical reaction like this it is neglected. The reaction (9.12) requires the direct contact of dissolved carbon and oxygen, i.e. they must be delivered to the reaction site and subsequently the product must be removed in the form of gaseous CO from the place of reaction. The mechanism of this process is related to the formation of a new phase – gaseous carbon monoxide. Therefore, the reaction can only occur at the liquid metal-other phase interfaces, where nuclei of the gaseous phase can form and grow to the critical sizes, enabling them to further grow and flow out. The interfaces between the liquid metal and the other phases present there, as pores in the refractory lining, solid nonmetallic inclusions, gas bubbles, etc., are surfaces like those. The kinetics of this reaction, or providing substrates and removing products, is strictly related to the temperature and movement of liquid metal and gas bubbles. If the metal bath is not stirred, the thermal movement of the metal is laminar, which means that the speed of movement of the reactants and reaction products, described by Fick's law, depends on the diffusion coefficient, the value of which increases as the temperature increases.

If oxygen gas is injected directly into the metal bath, it is more reasonable to adopt the so-called two-stage pattern for the carbon oxidation. Then, at the first stage, the oxidation of iron occurs, which is the main component of the liquid solution, i.e. the metal bath (according to the reaction 9.23). The iron oxide formed within the entire volume of the bath, being in the liquid phase but not dissolved in the iron solution, is reduced with carbon dissolved in the bath. At the same time, it means that iron oxide is an oxygen carrier, which causes the oxidation of carbon as well as of other components. The process of reducing iron oxide with carbon then proceeds according to the reaction:

$$(FeO) + [C] = [Fe] + \{CO\} \tag{9.26}$$

Some FeO is not involved in the reduction reaction and floats to the upper surface, passing into the slag.

The industrial experience shows that the kinetics of the reaction (9.26) is high, and at the same time, the injected oxygen jet introduces a large amount of kinetic energy. This energy changes the movement of the metal bath from a laminar to a highly turbulent. This results in turbulent swirls of liquid metal, which significantly increases the mass transport coefficient. At the same time, the pressure generated by the oxygen jet causes deep penetration into the liquid and its breaking into tiny bubbles, around which the iron oxidation and the formation of the oxide phase take place. At the interface of the oxide phase, the carbon dissolved in the metal is reduced and a new gaseous phase is created in the form of CO bubbles. The momentum given to the bubbles of oxygen and carbon monoxide as well as to fine metal and slag droplets causes the formation of a gas-metal-slag emulsion, which moves from the area of oxygen injection to the other places of the metal bath, and partially splashes above the surface. The experimental studies [13] show that the carbon content in metal droplets is always much lower than in the bath. The decarburized droplets, after falling into the bath, produce a decreasing carbon concentration gradient in the metal from the top to the bottom. The calculations show that the estimated share of carbon oxidation from the bath according to the mechanism described by the reactions (9.23) and (9.26) accounts for 35%–40% of the total decarburization. The other decarburization takes place in the volume of the metal bath, also mostly according to the above mechanism, but taking into account the turbulent transport of liquid metal with higher carbon contents from areas distant from the zone of direct influence of the jet of injected oxygen to the zone close to the oxygen jet with a significantly lower carbon content. The reaction occurs in the zones near the jet, but the transport of reactants is intensified by the turbulent diffusion. In general, the entire process of carbon oxidation is associated with the intensive stirring of the entire volume of the bath.

In steelmaking conditions, regardless of the method of oxygen supply (i.e. the adopted pattern of carbon oxidation), after reaching the temperature of approximately 1,843 K (1,570°C) by the metal bath, the rate of carbon oxidation reaction becomes so high that expanded gaseous CO bubbles occur along with vigorous stirring of the bath, called "boiling". The intense reaction of carbon oxidation is aimed at obtaining technical advantages: homogenization of the chemical composition and temperature in the entire volume of the bath, oxidation of undesirable elements (e.g. sulfur), removal of gases (hydrogen, nitrogen) from steel, heating the bath to the adequate temperature. Technical advantages of the oxidation process are particularly important if at least 0.3% of carbon is oxidized. Therefore, the carbon content in the charge is often increased by intentionally adding materials with a high carbon content (pig iron, coal, coke, etc.). In recent years, an inert gas has been used to increase the stirring intensity in the metal bath. This gas is blown through special porous plugs installed in the bottom of the furnace.

9.5.2 Sulfur Removal

The mechanism of sulfur removal reaction (desulfurization) in the steelmaking process is different than in the case of other metal bath components. It is a reaction of exchange of anions at the interface surface of metal and slag, according to the following formula:

$$[S]+\left(O^{2-}\right)=\left(S^{2-}\right)+[O] \tag{9.27}$$

Sulfur does not form any oxide or a complex anion in the slag. To chemically bond the forming sulfur anions, the slag should contain the "free cations" of the relevant elements. The most advantageous is calcium cation Ca^{2+}; therefore, lime is used in adequate amounts to form slag in the steelmaking process. Then, the reaction is presented as follows:

$$[S]+(\mathrm{CaO})=(\mathrm{CaS})+[\mathrm{O}] \tag{9.28}$$

The equilibrium constant of this reaction is:

$$K=\frac{a_{(\mathrm{CaS})}\cdot a_{[\mathrm{O}]}}{a_{(\mathrm{CaO})}\cdot a_{[\mathrm{S}]}}=\frac{a_{(\mathrm{CaS})}\cdot f_{\mathrm{O}}\cdot[\%\mathrm{O}]}{a_{(\mathrm{CaO})}\cdot f_{\mathrm{S}}\cdot[\%\mathrm{S}]} \tag{9.29}$$

where:

$a_{(\mathrm{CaS})}$ – activity of the formed sulfide in the slag,
$a_{(\mathrm{CaO})}$ – calcium oxide activity,
f_{O}, f_{S} – oxygen and sulfur activity coefficients,
[%O], [%S] – oxygen and sulfur concentrations in the metal bath.

The equilibrium sulfur content is:

$$[\%\mathrm{S}]=\frac{a_{(\mathrm{Cas})}\cdot f_{\mathrm{O}}\cdot[\%\mathrm{O}]}{K\cdot a_{(\mathrm{Cao})}\cdot f_{\mathrm{S}}} \tag{9.30}$$

For carbon contents lower than 0.3%, the activity coefficients can be assumed to be: $f_{\mathrm{O}}\approx 1$, $f_{\mathrm{S}}\approx 1$. Then the percentage share of sulfur in the metal is:

$$[\%\mathrm{S}]=\frac{1}{K}\frac{a_{(\mathrm{CaS})}}{a_{(\mathrm{CaO})}}\cdot[\%\mathrm{O}] \tag{9.31}$$

The dependence (9.31) shows that the extent of sulfur removal from the metal bath depends on the oxygen content in the bath, and the activity of calcium sulfide and calcium oxide in the slag. The temperature-dependent equilibrium constant of the reaction is an additional factor. The reduction of the sulfur content in the metal bath is influenced by the lower level of oxygen in it, the lower content of calcium sulfide, and the higher content of calcium oxide in the slag. At the same time, an increase in temperature has a positive effect on reducing the sulfur concentration.

Desulfurization of the metal bath with calcium oxide at a presence of carbon occurs according to the following reaction:

$$[\mathrm{S}]+(\mathrm{CaO})+[\mathrm{C}]=(\mathrm{CaS})+\{\mathrm{CO}\} \tag{9.32}$$

for which the equilibrium constant is:

$$K_1=K\cdot K_{\mathrm{C,O}}=\frac{a_{(\mathrm{Cas})}\cdot p_{\mathrm{CO}}}{a_{(\mathrm{CaO})}\cdot a_{[\mathrm{S}]}\cdot a_{[\mathrm{C}]}} \tag{9.33}$$

In the performed research conducted on desulfurization with the calcium oxide of a metal bath containing 0.3–1.8% C, it was found that within the temperature range of 1,773–1,873 K (1,500°C–1,600°C) the product of carbon and sulfur activity is:

$$a_{[S]} \cdot a_{[C]} = 0.0135 \pm 0.0015 \tag{9.34}$$

The value of the equilibrium constant of the deoxidation reaction with carbon at a temperature of 1,873 K (1,600°C) and a pressure of 0.102 MPa is 400, thus the product of carbon and oxygen activities under these conditions is equal to:

$$a_{[O]} \cdot a_{[C]} = 2.4 \cdot 10^{-3} \tag{9.35}$$

As a result of dividing the dependence (9.34) by (9.35), we obtain:

$$\frac{a_{[S]}}{a_{[O]}} = \frac{0.0135}{0.0024} = 5.6 \tag{9.36}$$

The dependence (9.36) shows that the activity of sulfur in a metal bath depends on the activity of oxygen in this bath. Assuming $f_O \approx 1$ and $f_S \approx 1$, we obtain the following dependence:

$$[\%S] \approx 5.6 \cdot [\%O] \tag{9.37}$$

The above dependence shows that when desulfurizing a metal bath with calcium oxide, it is not possible to obtain sulfur concentrations lower than six times the concentration of oxygen in the bath.

In industrial practice, the course of desulfurization is often assessed on the basis of the sulfur partition coefficient L_S, defined assuming the following course of the reaction (9.27):

$$L_S = \frac{(\%S)}{[\%S]} = K \cdot \frac{a_{O^{2-}}}{[\%C]} \cdot \frac{f_S}{\gamma S^{2-}} \tag{9.38}$$

where:

K – reaction equilibrium constant (9.27),
$a_{O^{2-}}$ – activity of O^{2-} in slag,
f_S – activity coefficient of S in liquid metal,
$\gamma_{S^{2-}}$ – activity coefficient of S^{2-} in slag,
$[\%O]$ – concentration of O in liquid metal.

From the definition of equation (9.27), it appears that the reduction of sulfur content in the liquid metal can be obtained by reducing the activity of anions S^{2-} and increasing the activity of anions O^{2-} in the slag. Both of these conditions can be met by reducing the content of acidic components of the slag. The sulfur anion has a larger diameter than the oxygen anion. The calcium cation has the largest size of all cations found in the traditional slag. Large sulfur anions are concentrated next to large calcium cations, and smaller oxygen anions are concentrated around iron and manganese cations. An increase in content of SiO_2 or P_2O_5 in the slag reduces the number of free anions O^{2-}. Therefore, the increase in the content of CaO in the slag helps reduce the activity of anions S^{2-}, and to increase the activity of the ions O^{2-}, which facilitates the passage and retention of sulfur in the slag [14].

It follows from the above considerations that the metal bath desulfurization process should be carried out at a high temperature with as low oxygen content as possible. During the oxidation period, the bath temperature is sufficient for the desulfurization reaction, while the oxygen content is unfavorable for this process. The oxidation, wherein the basic operation is injection of oxygen gas into the metal bath, causes a significant dissolution of oxygen molecules in the bath volume and an increase in its concentration to 500–1,000 ppm (0.050–0.100%). At the same time, the oxygen injected causes the iron to oxidize, according to the reaction (9.23), which produces a new liquid oxide phase FeO, with cations Fe^{2+} and anions O^{2-}. The effect of carbon oxidation during the oxidation period is an intensive turbulent metal movement and the formation of a metal-slag-gas emulsion. This emulsion includes metal droplets containing dissolved sulfur, and slag droplets containing oxygen anions but without sulfur anions. A system like this favors the course of the reaction (9.27), that is, the removal of sulfur from the metal and its transfer to the droplets of the slag phase, which "flow" to the surface of the bath, forming the slag. In the slag, sulfur anions can form stable compounds with the calcium cations present there. The above analysis shows that one of the basic conditions for removing sulfur from the metal bath is to ensure the presence of calcium cations in the slag, i.e. its appropriate basicity.

In industrial processes, the kinetics of the sulfur transition from liquid metal to liquid slag is very important. Mathematically, the description of this process can be derived on the basis of the kinetic law of mass action [14], which, when adapted to the reaction of oxygen and sulfur anions exchange between metal and slag, can be presented as follows:

$$-\frac{d[S]}{dt}\cdot\frac{G}{100F}=k_1\cdot[S]^x-k_2(S)^y \tag{9.39}$$

where:

G – metal weight, kg,

F – slag-metal interface area per unit of its volume, m^3,

k_1, k_2 – reaction rate constants for the transfer of sulfur from metal to slag and from slag to metal,

x, y – numbers characterizing the order of the reaction of sulfur transfer from metal to slag; when the process is controlled by the rate of a chemical reaction, they are equal to the stoichiometric coefficients.

In the initial desulfurization period, the sulfur content in the slag is very small – assuming that it is zero, the equation (9.39) can be simplified to the following form:

$$-\frac{d[S]}{dt}\cdot\frac{G}{100F}=k_1\cdot[S]^x \tag{9.40}$$

Then:

$$-\frac{d[S]}{[S]^x} = \frac{100F}{G} k_1 \cdot dt \tag{9.41}$$

After integrating equation (9.41) and assuming $x = 1$, we obtain:

$$\ln[S] - \ln[S]_0 = \frac{100F}{G} k_1 \cdot (t - t_0) \tag{9.42}$$

where:

$[S]_0$ – initial sulfur content in the metal bath (in time t_0).

The equation (9.42) describes the changes in the sulfur content of the metal as a function of time t, with respect to the initial sulfur content in the metal bath. The linear nature of changes is maintained only in the case of basic and neutral slags.

The factors controlling the kinetics of the desulfurization process include the following:

- transport of reagents in the volume of metal and slag, and in the border layers of these phases,
- anion exchange reaction at the interface of metal and slag.

The rate of reagent transport is the slowest stage of the kinetics. During the oxidation period, the intensive stirring of the bath and the formation of a metal-slag-gas emulsion accelerate the kinetics of the desulfurization reaction to such an extent that despite unfavorable conditions in terms of oxygen content in the metal, the sulfur removal reaction occurs to a large extent. This is confirmed in the industrial conditions.

9.5.3 Gas Removal

During the period of intensive oxidation of the metal bath, there are also favorable conditions for removing gases from the metal bath, especially hydrogen and nitrogen. These elements appear in the metal bath during the scrap meltdown in the EAF, when favorable conditions exist for their dissolution in the forming liquid ferrous solution. This is due to high temperatures in the area of electric arcs that favor gas ionization. In the gaseous atmosphere of the arc furnace reaction chamber, both nitrogen and water vapor exist, which can dissolve in the metal bath according to the following reactions:

$$\{H_2O\} = 2[H] + [O] \tag{9.43}$$

$$\{N_2\} = 2\,[N] \tag{9.44}$$

The amount of dissolved hydrogen or nitrogen in the bath is related to the partial pressure, respectively, $\sqrt{H_2O}$ or $\sqrt{N_2}$ in the gas atmosphere of the furnace, and this is numerically described by Siverts's law. After melting the charge and forming the metal bath, the contents of both gases present in the metal bath significantly exceed their levels in the finished steel. The hydrogen contained in steel always deteriorates its properties and it is, therefore, an undesirable component. Nitrogen also generally degrades the properties of steel, although there are grades for which nitrogen is considered an intentional alloying component. Current steel production practices provide for the removal of these gases from the metal bath during secondary refining in the ladle, mainly in vacuum devices. However, due to the cost, this is particularly true for grades with very low allowed gas contents. Manufacturing processes of common carbon steels do not provide for refining in vacuum devices, and the oxidation is the best period to reduce the hydrogen and nitrogen content in steel.

The mechanism of the process of removing hydrogen and nitrogen from the metal bath is associated with the formation of carbon monoxide gas bubbles in its volume and their turbulent movement, and outflow into the atmosphere of the furnace vessel. CO bubbles are treated as a medium that "absorbs" and "lifts" hydrogen and nitrogen from the bath. The first stage of the mechanism is the transport of dissolved [H] and [N] to the interfaces between the CO bubble and the liquid metal. Then, at the interface, the adsorption and transfer of hydrogen and nitrogen in the form of diatomic gas to the bubble volume occurs. The last stage is lifting the bubbles containing these gases above the bath to the furnace vessel atmosphere [5].

The kinetics of the degassing process depends on the speed of the metal movement relative to the carbon monoxide bubbles (flow turbulence), which is directly related to the rate of carbon oxidation. This concerns in particular the hydrogen removal, for which an experimental kinetic equation was determined:

$$-\frac{d[\%\mathrm{H}]}{dt} = -2.3 \cdot 10^4 [\%\mathrm{H}]^2 \cdot \frac{d[\%\mathrm{C}]}{dt} \tag{9.45}$$

The equation shows a relatively high hydrogen removal rate of up to 5 ppm/h. In the practical conditions, the obtained hydrogen content in the metal bath after oxidation period reaches values from 2 to 3 ppm [5].

The kinetics of nitrogen removal from the bath is described slightly differently [15]. It is assumed that it is related primarily to its oxygen content. In this case, the determined experimental kinetic equation (based on the law of mass action) has the following form:

$$\frac{d[\%\mathrm{N}]}{dt} = k \cdot \frac{A}{V} \cdot \frac{1}{[\%\mathrm{O}]}\left([\%\mathrm{N}]^* - [\%\mathrm{N}]\right) \tag{9.46}$$

where:

A – area of gas bubbles of CO, m^2,

V – metal bath volume, m^3,

$[\%N]^*$ – limit value of nitrogen solubility in the bath, for given conditions, %,

k – kinetic reaction factor of nitrogen removal.

The period of oxidation of the metal bath is a very important process stage because it prepares the liquid metal in terms of the homogeneity of the temperature field and homogenization of the chemical composition, and – which is particularly important – the content of undesirable and harmful elements. At later stages, it is not possible anymore. Although it is possible to melt the steel without the application of the oxidation period or with limited oxidation, but a practice like this requires very careful preparation of the charge, particularly in terms of the content of harmful, undesirable, and nonoxidizing components. Often, in the case of the production of common steels, there is no separate oxidation period in the production practice. Then, at the end of the meltdown period, the bath is oxidized, the extent of which depends only on the need to oxidize the undesirable elements. Operations homogenizing the chemical composition are carried out in secondary metallurgy facilities. In this case, the metal bath, which is a semi-product in the process of steelmaking in the arc furnace, usually has the following chemical composition: 0.05%–0.10% C; 0.08%–0.15% Mn; under 0.05% Si; under 0.02% P; under 0.02% S, and also very low levels of other alloying elements. The temperature usually does not exceed 1,923 K (1,650°C).

After the oxidation period, the metal bath does not contain undesirable elements, but it is characterized by high "oxygenation" (high oxygen content), no alloying elements, and the slag has a high oxidation potential. In classic technologies, the oxidation is followed by a period of refining the metal bath.

9.6 REFINING PERIOD

This process period is carried out only in the case of small, older design arc furnaces, in steelmaking shops not equipped with secondary metallurgy facilities [16]. The notion of refining includes a number of technical steps:

- deoxidation of the metal bath,
- formation of the refining slag,
- making up the chemical composition,
- obtaining an adequate metallurgical cleanness of the liquid metal,
- final deoxidation,
- heating the bath to the required temperature.

The primary goal of refining is to obtain the appropriate quality of the steel produced. It is understood as the chemical composition appropriate for the steel grade manufactured (in accordance with the applicable standards), and the appropriate metallurgical cleanness and the casting temperature established in the process assumptions.

The first stage of refining is the preliminary deoxidation of the metal bath. It is carried out by adding to the metal bath materials containing elements with a greater affinity for oxygen than iron. After the oxidation period, the metal bath contains several hundred *ppm* of dissolved oxygen. In order to be able to effectively make up the chemical composition of the metal bath, it is necessary to reduce the oxygen content in the bath to below 100 ppm. This objective is achieved by preliminary deoxidation, carried out with the use of so-called deoxidizers, the most common of which are ferrosilicon, ferroaluminum, sometimes ferromanganese or ferrosilicomanganese. The above-mentioned ferroalloys are used in a lump form and are added by throwing them into a metal bath through the main door.

The amounts of deoxidizers applied depend on the oxygen content in the metal bath, the types of steel produced, and the type of ferroalloy used. The oxygen content in the metal bath is bound by a hyperbolic relationship with its carbon content. At low carbon contents, the oxygen content is high; as the carbon content increases, the oxygen level decreases. Therefore, to deoxidize the metal bath when producing low-carbon steels, more deoxidizers must be used, and less with high-carbon ones. The consumption of deoxidizers is also related to the adopted practice of steel refining: under one or two slags.

In the case of the production of higher-quality structural steels as well as medium- and high-alloy steels, refining practices under two slags are used. Then, the refining period begins from the formation of the refining slag with appropriate properties, the most important of which is a low content of iron oxides. Therefore, the first operation is to remove the slag from the furnace after the oxidation period, as it contains large amounts of FeO. Only after that, the deoxidation of the bath is started, using 20% ferroaluminum, in the amount of 0.3–1.5 kg/t of metal charge, depending on the carbon content in the bath. Simultaneously with FeAl, ferrosilicon and ferromanganese, rarely ferrosilicomanganese, are added in amounts depending on the content of silicon and manganese in the steel grade made, respectively. The minimum amounts of FeSi and FeMn are 1–2 kg/t of the metallic charge. After a few minutes, the alloying elements are introduced in the amounts depending on the allowed content of the relevant elements in the steel grade produced. Only then, the refining slag formation begin.

Lime in the amount of approximately 2 kg/t of metallic charge is used as a slag-forming additive. The lime is dissolved in a small part of the slag remaining after oxidation; together with the flowing out of nonmetallic inclusions, and the worn-out refractory lining of the furnace, it forms a refining slag containing up to several percent of FeO. This amount of iron oxides means that the slag has an oxidizing character and needs to be deoxidized. For this purpose, deoxidizing agents are applied onto the slag surface in the form of finely ground graphite electrode waste, sometimes ferrosilicon or aluminum. This results in a reduction reaction of the oxides contained in the slag:

$$(\text{FeO}) + \text{C}_{st} = [\text{Fe}] + \{\text{CO}\} \tag{9.47}$$

$$(\text{MnO}) + \text{C}_{st} = [\text{Mn}] + \{\text{CO}\} \tag{9.48}$$

$$(\text{CrO}) + \text{C}_{st} = [\text{Cr}] + \{\text{CO}\} \tag{9.49}$$

$$(\text{P}_2\text{O}_5) + 5\text{C}_{st} = 2[\text{P}] + 5\{\text{CO}\} \tag{9.50}$$

As a result of the deoxidizing reaction, the iron oxides in the slag are reduced to less than 1%, the manganese and chromium oxides to less than 0.5%, and the phosphorus is reduced by about 0.01%. Slag deoxidized this way has the following chemical composition: 50%–60% CaO, 15%–20% SiO_2, 5%–10% MgO, and 3%–6% Al_2O_3. Due to its appearance, such slag is called white and has good desulfurizing properties. The desulfurization process proceeds according to the reaction:

$$[S]+(O^{2-})+C_{st}=(S^{2-})+\{CO\} \tag{9.51}$$

However, during the refining period, during deoxidation, the mixing of bath and slag is limited, and the movement of components within their volumes occurs by diffusion only. The reaction (9.51) takes place at the slag-metal and the slag-floating carbon molecules interfaces and, therefore, the desulfurization rates obtained in such conditions do not exceed 20%–40%.

The above-described stages of preoxidation, silicon and manganese content replenishment, and slag deoxidation take about 15–20 min in total. After that, a metal sample is taken from the bath and its chemical composition is analyzed. Based on the obtained result, the differences between the contents of individual elements in the bath and the contents required by the standards for the steel grade made are calculated. On this basis, the amount of additions necessary to make up the chemical composition of the bath before tapping is determined. The additives are ferroalloys, which bring the appropriate elements into the bath. They are added to the furnace in a lump form by throwing them through the main door. After waiting several minutes, until the ferroalloys dissolve, the final metal sample is taken in order to analyze the chemical composition. If the analysis result complies with the process assumptions, the metal temperature is measured. If the right temperature is obtained (according to the process assumptions, depending on the steel grade), the liquid metal bath is ready to tap.

In the case of making common structural carbon or low-alloy steels, the refining practice under a single slag can be adopted, which includes only deoxidation, making up the chemical composition and tapping. The deoxidation is then carried out in a furnace using 20% FeAl, the addition of which is dependent on the carbon content of the bath. For contents below 0.1%C, the addition of ferroaluminum is about 5 kg/t of metallic charge; for contents in the range of 0.1%–0.2%C, this addition is 2–4 kg; and for higher carbon contents, the addition of FeAl is about 1.5 kg. At the same time, ferrosilicon and ferromanganese may be added in quantities depending on the content of silicon and manganese in the steel grade made. In the steelmaking practice, two types of ferrosilicon are used, with a silicon content of 45% or 75% Si. FeSi 45%, characterized by a higher specific gravity, which is easier to dissolve in a liquid steel bath, is added to the furnace. The commercial ferromanganese contains about 65% Mn, but grades with different carbon content are produced: high carbon (about 6.5%C), medium carbon (about 3.5%C), and low carbon (below 1%C). The price of FeMn is related to its carbon content, the lower the C content, the higher the price. For this reason, for steels above 0.15% C, high carbon FeMn is used, and for low carbon steels low carbon, FeMn is used. As a rule, steels made using the practice under a single slag do not require the addition of other alloying components, but if it is necessary, they should also be supplemented at the same time.

After waiting several minutes, until the ferroalloys dissolve, the final metal sample is taken in order to analyze its chemical composition. If the analysis result complies with the process assumptions, the metal temperature is measured. When the temperature is correct, the liquid metal bath is ready to be tapped.

The process of dissolving the added ferroalloys is related to the deoxidation of steel. The whole dissolution process is complicated, involving successively the phenomena of mixing the bath, their dissolution as such, as well as resulting from oxidation: nucleation and growth of inclusions, and their flowing out toward the top surface. Convective movements of the bath streams play an important role in this process, which determine the flowing out and removal of the formed nonmetallic inclusions. During the refining period of the metal bath, the movements of the liquid are small compared to the oxidation period. They only result from the temperature gradients present in the volume of the liquid metal. The thermal energy from the electric arcs is provided to the upper surface of the bath, and in this region, the highest temperatures occur, while in the lower zones, the bath temperatures are lower. For this reason, the dissolution rate of ferroalloys is not too high, and the supersaturation with the elements contained in the added materials appears in the bath areas around their pieces. As a result, in these areas it becomes possible to homogeneously nucleate the reaction products of the dissolved elements with oxygen, which is also dissolved in the bath. The occurring deoxidation reactions are similar as in the case of the oxidation of the metal bath:

$$[O]+[Al]=(Al_2O_3) \tag{9.52}$$

$$[O]+[Si]=(SiO_2) \tag{9.53}$$

$$[O]+[Mn]=(MnO) \tag{9.54}$$

It is only noteworthy that, during the oxidation period, dissolved elements are present in the bath, brought in with the charge, and oxygen is supplied from the outside to oxidize them. However, during the deoxidation stage, the bath contains dissolved oxygen, and ferroalloys containing elements (with a high affinity for oxygen) reacting with it are added to reduce its content. From the chemical point of view, these are the same oxidation reactions, but the purpose of initiating such reactions is completely different. In the oxidation period, the reactions are called oxidizing, and in the refining period, they are called deoxidizing.

The products of both the reaction (9.23)–(9.25), and the reaction (9.52)–(9.54) is a new phase – the oxide phase – that is liquid (FeO only) or solid for the other oxides. The reactions occurring during the oxidation period are heterogeneous, taking place mainly at the interfaces of gas bubbles of CO and with their turbulent movement they quickly move toward the top of the bath. However, the reactions occurring during the refining stage have a completely different nature and mechanism. The nuclei of the reaction products may appear homogeneously in the bath volume due to numerous supersaturations of the concentration of elements introduced with ferroalloys. In this case, the activation energy of nuclei formation, depending on the magnitude of the interfacial tension between the liquid iron solution and the formed oxide phase, significantly decreases [14,17]. This concerns in particular the forming nuclei of Al_2O_3 [18].

Under the conditions of the steelmaking process, these nuclei are in the solid phase. The test results indicate that in a ferrous solution containing 0.05% dissolved oxygen, approximately 10^5 of oxygen nuclei may occur in 1 cm^3 of the bath.

The resulting solid nuclei are the site of further deoxidation reactions, and the enlargement of the formed nonmetallic inclusions. At the same time, the inclusions flow in a laminar way upward at a speed that can be calculated from the Stokes law:

$$v = \frac{2gr^2(\gamma_m - \gamma_W)}{9\eta} \tag{9.55}$$

where:

v – the speed of the outflow of a particle (nonmetallic inclusion), m/s,
g – constant of gravity, m/s^2,
r – particle radius (nonmetallic inclusion), m^2,
η – bath dynamic viscosity coefficient, Pa·s,
γ_m – specific gravity of the bath, N/m^3,
γ_w – the specific gravity of the particle (nonmetallic inclusion), N/m^3.

Formula (9.55) applies to spherical particles, the diameter of which does not exceed 100 μm. Experimental studies show that under steelmaking conditions, the deoxidation products do not exceed this dimension [18]. The analysis of the formula (9.55) shows that the following factors increase the flow rate in laminar conditions: a larger particle diameter, a greater difference of metal and particle specific gravities, and a lower viscosity.

From the point of view of the outflow speed, the most advantageous are the inclusions of Al_2O_3, which tend to coagulate and form larger particles. Significant acceleration of the flow rate is achieved by stirring the metal bath, which is sometimes used practically by blowing inert gases (argon or nitrogen) into its volume. For this purpose, special porous plugs are placed in the furnace bottom. The technically favorable influence on the conditions of inclusion outflow is obtained by the simultaneous use of several ferroalloys or materials known as complex alloys for deoxidation. This is because complex nonmetallic inclusions, with complex chemical constitution form, such as aluminosilicates or manganese silicates. These types of inclusions flow out of the bath more easily and faster; they also have a more favorable effect on their so-called morphology, which has a less detrimental effect on the properties of the finished steel.

The initial deoxidation time is related to the amount of oxygen removed, and the types of deoxidizers used. It is also related to the grade of steel produced, especially the carbon content. Taking into account the time needed for the deoxidation products to flow out, the deoxidation time is largely determined by the mixing capabilities of the metal bath. In total, this stage of the refining period lasts from several to 30 min.

After deoxidizing the metal bath, the chemical composition is made up. The purpose of this stage is to obtain the content of individual components, in accordance with the process assumptions for a given grade of steel produced. This step begins with taking a metal sample to analyze the chemical composition of the metal bath. Based on the result of the analysis, the amount of supplementary additions needed is calculated. Ferroalloys bringing in the appropriate elements making up the chemical composition of the bath are added. When calculating the amount of an appropriate ferroalloy, the content of a given element in this ferroalloy should be taken into account. If, for example, the calculations show that 100 kg of manganese should be supplemented in the metal bath, and we intend to use ferromanganese containing 70% Mn, the addition of FeMn should be $100\ \text{kg} \cdot \frac{100\%}{70\%} = 143\ \text{kg}$. During the calculations, the so-called yield of a given element should also be taken into account, i.e. the share of the amount of the element that will dissolve and remain in the metal bath until the solidification into the finished steel, in relation to the amount of the element introduced with the ferroalloy. The term "melting loss" is sometimes used to denote the proportion of the element that will dissolve and will not remain (will oxidize) in the metal bath until it gets solidified into the final steel, compared to the amount of the element introduced with the ferroalloy.

There are no absolute rules for the order in which ferroalloys are entered into the bath. However, there are some general rules arising primarily from the chemical affinity of the elements added for oxygen, in the context of the affinity of iron for oxygen. Materials bringing elements with a lower affinity for oxygen than iron can be added at any time during a steelmaking process. They are usually introduced at the beginning of the manufacturing process, often simultaneously with the scrap loaded in the charging basket. Such materials include nickel, molybdenum, tungsten, and cobalt. Nickel and cobalt are added in the form of so-called technical metals with more than 99% Ni or Co, respectively. An additional argument for the early addition of nickel to the steelmaking process is the natural, relatively high hydrogen content of nickel-bearing materials. In this case, there is enough time to remove the hydrogen introduced and transferred into the bath, especially during the oxidation period when the conditions for its removal are favorable. Molybdenum is added in the form of ferromolybdenum containing more than 58% molybdenum. Tungsten is added in the form of ferrotungsten containing about 70% tungsten. It can be assumed that the yield of nickel, cobalt, or molybdenum is practically 100%. The yield of tungsten is slightly lower, about 90%. By contrast, the addition of tungsten at the start of the process is justified by another reason; it stems from a very high melting point of tungsten (3,695 K (3,422°C)) and its slow dissolving in the metal bath.

Elements with a chemical affinity for oxygen greater or similar to iron are added during the refining period. At the beginning of refining, manganese and chromium are introduced after initial deoxidation. Both the elements are introduced by using ferroalloys: ferromanganese or ferrochromium. The affinity for oxygen of both manganese and chromium is slightly higher than that of iron and, therefore, these elements oxidize to a small extent. The yield of both elements exceeds 90%. Note that three types of ferrochrome are commercially available, differing in their carbon content. The most commonly used, due to its low price, is the high-carbon FeCr, containing over 6% carbon. Carbon added with ferrochrome also dissolves in the metal bath, which further increases its content in the liquid metal. As it is not possible to oxidize carbon during the refining period, and until the end of the melting process, the extent of carbon oxidation during the oxidation period should be appropriately anticipated.

Elements with a significantly higher chemical affinity for oxygen than iron are added at the end of the refining period, only after the final deoxidation of the bath. This applies, for instance, to titanium or vanadium. Both elements are added using ferroalloys, respectively, ferrotitanium or ferrovanadium. The affinity for oxygen of both titanium and vanadium is much higher than that of iron and, therefore, these elements oxidize to a significant extent in the metal bath. The yield of both elements is approximately 50%.

For the production of common carbon steels, the content of alloying elements is low and, therefore, the need for added components is also low. As a rule, manganese and silicon are made up to a content of about 0.2%–0.5%. This means adding 3–8 kg of ferroalloys per 1 ton of the bath. When high-alloy steels are produced, the share of alloying elements can reach 20%, and sometimes even more. Then, the amounts of added ferroalloys are significantly higher; they can reach 300 kg/1 metric ton of bath. Small amounts of ferroalloys dissolve relatively quickly and easily in the bath and do not change its heat balance. On the other hand, large amounts of ferroalloys added, reaching up to 20%–30% of the bath weight, require a long time to dissolve and significantly lower the melt temperature. In this case, the ferroalloys are heated up before being loaded into the furnace (to a temperature of about 1,073 K (800°C)), the maximum power of the arcs is turned on and the feeding can be divided into several portions.

After adding additions to make up the chemical composition of the bath during the refining period, we should wait some time, up to a dozen or so minutes because the added materials must have time to dissolve. After waiting, a metal sample should be taken in order to analyze the chemical composition and check whether the obtained values comply with the process assumptions. If the result shows deficiencies of an element, a correction must be made. It should be taken into account that obtaining a result with exceeding one or a few elements in the bath, in relation to the acceptable upper limits in the steel grade made, is a serious engineering error and makes such a heat incorrect. This is due to the inability to remove the excess of this element or a few elements in this heat, and the produced steel is rejected.

After obtaining the correct chemical composition of the bath, its temperature is measured. The temperature should be consistent with the process assumptions that make its value dependent on the chemical composition of the liquid metal. The measured temperature should be 50°C–100°C above the melting point of the metal. The melting point of elemental iron is 1,811 K (1,538°C). The metal bath, which is a solution of iron and alloy additives, is characterized by a lower melting point because adding the alloying elements decreases its value.

The last operation of the refining period is the final deoxidation of the metal bath. This deoxidation is performed using a so-called strong deoxidizer, an element that has a much higher affinity for oxygen than iron. The most commonly used materials are aluminum, ferroaluminum, or 75% ferrosilicon. These materials are added in the form of lumps, in the amount of 0.2%–0.5% of the bath weight. This is followed by pouring the liquid steel from the furnace, an operation technically known as the "tapping". In total, the refining period lasts from 0.5 to 1 h.

The above-described principles of refining the metal bath indicate the general rules of conducting this process period, which determines the quality of the steel produced. Many steelmaking shops have specific production and cost conditions that lead to the introduction of certain modifications and changes in relation to the standard process solutions of the refining period. An example of such a modification is the application of a solution involving the introduction of strong deoxidizers at the beginning of refining, immediately after removing the oxidizing slag. It is only after the intensive deoxidation of the metal bath that new refining slag should be formed. Then, the operation of deoxidation by the slag is marginalized, and the entire refining period is shortened. However, this solution may not always be applied due to the changed method of removing nonmetallic inclusions, which affects the metallurgical cleanness of the steel manufactured.

As mentioned at the beginning of the discussion of the steelmaking practice in the EAF, the refining period is only used for small furnaces that produce alloy steels. In modern steelworks equipped with large furnaces and secondary metallurgy facilities, the refining period in the furnace is not carried out. Then, immediately after melting the charge down, with a limited oxidation period already carried out at the end of meltdown, the metal bath is tapped. All of the refining processes are carried out in the ladle at the secondary metallurgy stations.

9.7 TAPPING

The final process stage of steel production in the EAF is the tapping of the metal bath, which already has the chemical composition of the final steel. Steel is tapped into the casting ladle, placed under the furnace. In practice, two tapping methods are used: through a pouring spout or through a taphole.

In the case of a pouring spout, the tapping operation begins with placing the casting ladle under the furnace, suspended on an overhead crane. The ladle should be positioned so that the end of the pouring spout is in a place above the ladle so that the flowing steel stream reaches its central part. It is important that the stream does not flow directly onto the ceramic sidewall of the ladle because this material is then leached out and the steel is contaminated with nonmetallic inclusions, and premature wear of the ladle occurs. Then, the taphole is pierced, which was filled with a refractory mass for the time of melting. Before starting the tapping, make sure that the entire runner is clean of dirt. After such preparation, the furnace is tilted toward the runner, until a stream of liquid metal appears. From now on, the furnace tilt angle should be controlled and the ladle should be positioned in such a way that the stream of flowing metal is "compact and continuous". This is very important from the point of view of the metallurgical cleanness of the steel manufactured. From the time of the liquid metal flowing out of the taphole, until it flows into the volume of the liquid metal in the ladle, the stream is in direct contact with the atmospheric oxygen. The liquid ferrous solution is characterized by a very high tendency to absorb oxygen, and the extent of this phenomenon depends on the size of the contact surface. The more compact the stream, the shorter its flow path and the lower the possibility of oxygen absorption into the liquid metal. An increase in the oxygen content in the bath results in the reaction of the oxidation of iron, as well as silicon and aluminum, resulting in the formation of impurities in the form of nonmetallic inclusions. It is also possible to absorb nitrogen, and in special cases also hydrogen into the steel. The time of tapping the metal into the ladle is related to the weight of the heat and the size of the taphole. The average time is 5–7 min.

In the case of a taphole in the bottom, the casting ladle is placed on a transport platform that moves along tracks laid under the furnace, perpendicular to the longer side of the furnace hall. The ladle should be placed under the furnace, centered on the taphole. In this method, there is no need to maneuver the position of the ladle. The liquid metal is poured vertically from the taphole directly into the ladle. The diameter of the taphole is adjusted to the size of the furnace so that the tapping lasts no longer than 3 min. After the ladle is positioned, the taphole, which was filled with refractory mass held by a closing flap from the bottom, becomes unblocked. After opening the flap, the refractory mass flows out under its own weight. This method of tapping guarantees much more favorable conditions for the protection of the liquid steel against "oxygenation", thanks to the more compact stream and its shorter path from the hole to the volume of the ladle.

Regardless of the method, care must be taken to prevent the furnace slag from getting into the ladle during tapping. An operation like this is not easy, as there is always the phenomenon of the liquid swirl when poured out of a container. In case of liquid metal in the furnace, with its surface covered with the liquid slag, the swirling phenomenon also appears, and at the end of pouring, the steel and slag are simultaneously flowing out. There are many patented designs, including practical solutions for finding the right moment to end the tapping, so that all of the liquid metal is poured out and all the slag remains in the furnace. However, none of the existing proposals is satisfactory enough. Two methods are used in the production practice.

In the first one, tapping to the ladle is ended before all of the liquid metal has been poured. Then, part of the steel, including the slag, remains in the furnace, as a result of which the production yield of steel in relation to the loaded charge materials decreases, which translates into a lower economic effect of the steelmaking shop operation. In the other method, metal is deliberately tapped with the slag. For the practice with the full refining period, the furnace slag is well deoxygenated and its presence in the ladle is recommended. It has favorable properties to protect the steel against oxygenation during its stay in the ladle from tapping to casting.

For a practice without refining, the furnace slag that has entered the ladle is corrected with the appropriate additions of slag-forming materials in order to improve its properties. Often in the production practice, indirect methods are used, in which a small part of the metal remains in the furnace, and a small part of the slag passes into the ladle. The selection of the method is individually adjusted to the production range, furnace capacity, secondary metallurgy scope, etc.

The practice where some of the metal is left in the furnace during tapping is quite common in the production of common carbon steels, in large-capacity furnaces. It is called the "hot heel" practice. Then, about 5% (sometimes up to 10%) of the weight of the metal bath is intentionally left in the furnace. The tapped metal certainly does not contain slag, and the liquid metal left in furnace facilitates the scrap meltdown and the formation of the slag for the next heat. However, this solution also has some disadvantages: it reduces the production efficiency of a given furnace unit, and it also makes it difficult to control the technical condition of the bottom after each heat.

After tapping the steel, the furnace is tilted toward the main door and the slag is tapped. The slag flows through the main door to a special pit under the furnace or to a cast iron slag ladle. In both cases, it is periodically evacuated from the pit or after the ladle is full to landfills for a later use.

During tapping to the ladle, part of the materials for the final deoxidation of the metal bath and slag-forming materials are added to the ladle. After tapping completion, the ladle with liquid metal is transported to the casting bay or to the secondary metallurgy stations.

9.8 ROLE OF SLAG

Generally, these three immiscible phases take part in steelmaking processes: liquid metal, liquid slag, and gas. The steel production process, at the stages with the presence of liquid metal, is always conducted with the participation of slag. Liquid slag is generally a multicomponent system of chemical compounds primarily containing oxides as well as silicates, phosphates, sulfides, etc. In the classic thermodynamic perspective, it is assumed that slag is a liquid solution, primarily consisting of oxide molecules (CaO, SiO_2, FeO, etc.) and other compounds as well as various chemical systems and eutectics forming between them [10]. This molecular theory describes slag in a very simplified manner and does not explain many physicochemical effects occurring with the slag participation. However, it is still frequently used for process considerations and simplified computing in industrial conditions. A more developed theory assumes that slag is an ionic solution wherein the behavior of the components in metallurgical processes is characterized by their thermodynamic activities.

The ionic theory assumes that the liquid slag contains:

- metal cations, such as Ca^{2+}, Mn^{2+}, Fe^{2+}, Fe^{3+}, Mg^{2+}, Al^{3+},
- simple anions, such as O^{2-}, S^{2-}, F^{1--},
- complex anions, such as SiO_3^{3-}, AlO_2^{1-}, FeO_2^{1-}, PO_4^{3-}.

Based on the adopted theory of slag, the processes occurring within the slag volume as well as the ones occurring at the interface between the slag and the metal or the slag and the gas phase can be analyzed. This applies to the following phenomena and processes:

- thermodynamics of reactions (e.g. determining the direction of reaction course),
- kinetics of reactions,
- scope of reactions (e.g. quantitative determination of the capabilities of "absorbing" sulfur or phosphorus).

In metallurgical practice, slags are usually determined by specifying their chemical constitution expressed either by the weight percentage or mole fractions. The most frequently determined slag constituents include CaO, SiO_2, MgO, MnO, FeO, P_2O_5, Al_2O_3, S, and Fe_2O_3. In some slags, subject to the needs, other constituents are also established and specified, for instance, TiO_2, Cr_2O_3, V_2O_3, Na_2O, K_2O, CaC_2, etc.

The knowledge of the slag structure, determining the slags' behavior in the metallurgical process conditions, is essential for the description of the occurring reactions and accompanying phenomena, which is necessary to correctly carry out process operations, and to optimally design new engineering procedures. The features of the slag structure are described by chemical or physical properties. In terms of the chemical nature, slags can be divided into two groups, namely:

1. Acid slags, which have an excess of acid oxides, such as SiO_2, P_2O_5, Fe_2O_3, Cr_2O_3, V_2O_3, and TiO_2, compared to basic oxides, such as CaO, MgO, MnO, and FeO.
2. Basic slags that have more basic oxides than acid oxides.

Alumina oxide is considered by many authors [18] as a constituent with amphoteric properties, i.e. in an acid slag, it can show the nature of a basic component, whereas in a basic slag, it can behave as an acid component. In practice, this division is stricter, by introducing an indicator called slag basicity. For steelmaking slags, the slag basicity is defined in its simplest form as the quotient of the percentage share of CaO and SiO_2:

$$V = \frac{(\%CaO)}{(\%SiO_2)} \tag{9.56}$$

Sometimes, a definition taking into account the content of P_2O_5 is applied:

$$V = \frac{(\%CaO)}{(\%SiO_2)+(\%P_2O_5)} \tag{9.57}$$

Other methods of defining the steelmaking slag basicity are also encountered. There is, for instance, a definition similar to (9.56) or (9.57), where the mole fractions of these compounds are used instead of weight percentage. Another example is the optical basicity, the definition of which takes into account mole fractions, the number of oxygen atoms, and the basicity of molecules present in the slag. Such definitions are applicable in scientific research. In industrial practice, the notion of basicity defined with the formula (9.56) is still applied, as this describes the chemical nature of slag in a very simple manner. The slag basicity in a steelmaking process directly indicates its capabilities to "absorb" and permanently bind products of reactions of phosphorus and sulfur removal from the metal bath – being among the most important engineering operations during steel production.

The second important chemical property of slags is called oxidizing capacity. Steelmaking processes have an oxidizing nature and, therefore, the contents of oxygen dissolved in the melt, and oxide anions in the slag are crucial for the course of oxidizing reactions. In the metallurgical practice, the oxidizing capacity means the capability of the metal bath and the slag to oxidize the constituents that have an affinity with oxygen higher than with iron, as the main component of steel. Primarily, elements with a variable cation valence influence the oxygen behavior in a steelmaking process. Iron is the most important element, which in steelmaking conditions can occur in slag as Fe^{2+} and Fe^{3+}. Manganese is also an element like this, but for steel production, due to its much lower content compared to iron, its significance is marginal. The slag oxidizing capacity is characterized by a parameter called the oxygen distribution coefficient, being the quotient of iron oxide activity in the slag and activity of oxygen dissolved in the metal bath.

$$L_{FeO} = \frac{a_{(FeO)}}{[O]} \tag{9.58}$$

According to the Nernst distribution law, the ratio written by the equation (9.58) is constant at a given temperature. This means that in slags containing more iron oxide, more oxygen dissolved in the melt is in equilibrium, or slags like these have a higher oxidizing capacity. The ratio of cations Fe^{3+} to Fe^{2+} is also important; the higher it is, the higher the oxidizing capacity will be. The basicity is also advantageous because, as the basicity increases, the content of oxygen anions O^{2-} increases, which enhances the slag oxidizing capacity. The oxidizing capacity is also characterized, in a sense, by the oxygen content in the metal bath.

The relevant physical properties of slag include density, viscosity, surface tension, and electrical and thermal conductivity. All of these properties depend on the chemistry of slags and the temperature. The density (specific gravity) is related to the structure of slags, indicating its ordered state and the existence or lack of interactions between its constituents. It also suggests the form of constituents, primarily the type of the formed complex anions. The density of steelmaking slags ranges from 2.8 to 3.5 g/cm^3.

Viscosity is the most important physical property of slags, accountable for the rate of mass exchange in the system. Viscosity is a feature characterizing the internal friction arising from shifting the fluid layers against each other during their flow. In steelmaking conditions, it concerns liquid metal and slag moving against each other. The temperature has a very high impact on the viscosity value. At steelmaking process temperatures, the liquid steel viscosity does not exceed 0.01 Pa·s, whereas the viscosities of liquid slags at these temperatures reach values at least ten times higher, of 0.1–3.0 Pa·s. Therefore, viscosity is the decisive factor influencing the kinetics of chemical reactions between the metal and slag. In basic steelmaking slags, the chemical composition determines their viscosity. Slags containing large amounts of iron oxides feature a medium viscosity. Dissolving silica in iron oxides significantly reduces, whereas dissolving calcium oxide increases viscosity. Al_2O_3 is a component, which reduces viscosity in basic slags.

The surface tension and related interfacial tensions influence the effect of slag foaming, their impact on the refractory lining, the course of chemical reactions, and the assimilation of nonmetallic inclusions. The surface tension is measured by the work that must be performed to create a unit surface area of a liquid. The influence of temperature and chemical constitution on the surface tension of basic slags is much lower than the influence of these parameters on the viscosity. The values of surface tension range from 0.30 to 0.45 J/m^2. For process reasons, the slag surface tension should be low, while the interfacial tension should be high, as it is then that slags show good refining properties and easily detach from the metal bath.

The heat conductivity of the slag is important in the aspect of heat exchange with the metal bath. The values of thermal conductivity coefficient of slags range from 2.3 to 3.6 J/(m·s·K) and are lower than the thermal conductivity coefficient of the metal bath by one order of magnitude (20–23 J/(m·s·K)). The effective thermal conductivity coefficient of foamy slag is from 4 to 7 J/(m·s·K). Intensive stirring, both of the slag and the metal bath, significantly increases the thermal conductivity coefficient. The slag heat capacity shows a significant dependence on temperature and practically does not depend on the slag chemical constitution.

The electrical conductivity of slags is relevant for testing not only their structure but also their practice and production. It concerns in particular the conditions of arcing. The presence of constituents prone to ionization in the liquid slag solution creates better conditions for stable arcing and increases the effectiveness of heat transfer from the arc to the metal. The electrical conductivity of slags increases as the contents of iron oxides and calcium oxide, as well as the temperature, increase.

Slag is indeed a by-product (auxiliary product), but it plays a very important technological role in the metallurgical production process. The most important functions of slag are as follows:

- to protect the metal bath against the impact of oxygen from the furnace atmosphere,
- to absorb products of oxidation of metal bath impurities, including phosphorus and sulfur,
- to assimilate nonmetallic inclusions.

Appropriate physical and chemical properties of slag largely determine the quality of the steel produced.

9.8.1 Functions and Properties of Slag during the Steelmaking Process in an Electric Arc Furnace

In steelmaking processes, slag forms already at the beginning of the melting. Its chemical constitution and properties continuously change during melting. This is caused by:

- oxidation of metal bath impurities,
- dissolution of slag-forming materials,
- erosion of refractory lining,
- temperature changes.

The following oxides are the fundamental constituents of the steelmaking furnace slag: CaO, SiO_2, MgO, MnO, FeO, Fe_2O_3, Al_2O_3, and P_2O_5. The share of individual constituents varies and continuously changes during the manufacturing process. The steelmaking slag, regardless of the process stage and its chemical constitution, always has oxidizing properties and contains relatively high amounts of sulfur and phosphorus.

In an EAF, in the first phase of charge meltdown, iron from the charge scrap is oxidized, resulting in the formation of primary slag, containing mainly iron oxides (called ferruginous slag). The main constituent of ferruginous slags is FeO, with the melting temperature of 1,642 K (1,369°C). The share of Fe_2O_3, with the melting temperature of 1,839 K (1,566°C), is small – not exceeding 2%. In addition, already in the initial phase of the meltdown, silicon is oxidized, and the forming silica SiO_2 dissolves in FeO, forming a solution with an even lower melting temperature. At an about 20% share of SiO_2, the temperature of the formed eutectic $2FeO{\cdot}SiO_2$ (fayalite) drops down to about 1,473 K (1,200°C). Exact data can be read from the diagram shown in Figure 9.9.

You can read from the diagram as to which phases are present in the system as a function of temperature. The share of FeO and SiO_2 is marked on the *X*-axis. On the left-hand side ("0"), there is only FeO in the system, whereas on the right-hand side ("1"), there is only SiO_2. Intermediate values correspond to the changing share of both oxides, assuming that their sum is 1. As can be seen from the diagram, at the SiO_2 share between 0.2 and 0.4, and at a temperature of 1,473 K (1,200°C), the system comprises liquid fayalite and metallic iron.

The lime added with the charge, other oxidation product, and wearing refractories dissolve in the formed slag consisting of iron and silicon oxides. At the end of the meltdown period, the chemical constitution of slag is as follows:

CaO: 25%–35%,
SiO_2: 20%–30%,
FeO: 25%–40%,
MgO: 5%–10%,
MnO, Al_2O_3, Cr_2O_3, P_2O_5 < 2%.

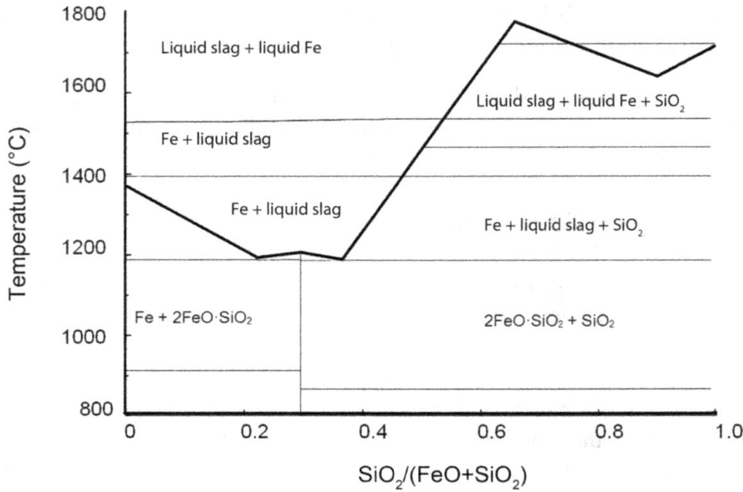

FIGURE 9.9 Equilibrium diagram FeO – SiO_2 versus temperature (Author's work).

The most important functions of the slag during the meltdown period include protection from the oxidizing impact of the gaseous atmosphere in the furnace, dissolution of products of oxidation reactions of the metal bath constituents (including sulfur and phosphorus), and the obtaining of properties favorable to the foaming process. Then, the slag features an oxidizing nature, a low basicity, a relatively low melting temperature (approximately 1,573 K (1,300°C)), and a medium viscosity. The purpose of the optimization of process conditions in the initial period of the slag formation is to quickly initiate its development. The application of good quality lime (high CaO content, low moisture), an appropriate grain size (preferably from 5 to 40 mm), and an effective method of introducing (lance injection) contribute to the accomplishment of this objective.

After the charge meltdown, further stages of the melting process depend on the type of practice. The functions and properties of the slag are also different. For the "full" procedure, with the oxidation and refining periods, the slag after meltdown stays in the furnace, and the oxidation period is carried out without intentional slag forming. Changes in the slag chemistry result from the dissolution of the products of oxidation of melt constituents, and changes in the physical and chemical properties related to the increase in temperature. The oxidation of phosphorus and sulfur from the metal bath should characterize this process period. The scope of oxidation of these elements depends on their contents in the metal bath after meltdown and the needs arising from the requirements concerning the steel produced. The oxidation products of these reactions float to the slag and dissolve. For the correct process of "absorption" of the phosphorus and sulfur oxidation products by the slag, the slag should have an appropriate basicity, and more precisely a sufficient amount of oxygen anions O^{2-}. Only then, the desulfurization reaction can occur, and products of phosphorus oxidation can be bound to stable compounds. Therefore, in those cases with a large scope of phosphorus and sulfur oxidation, to ensure a sufficient amount of oxygen anions, some amount of slag is racked off from the furnace and adequate amounts of lime are added. Thereby, a part of the phosphorus and sulfur compounds contained in the slag is removed with the slag, and the formed new slag, with a high basicity can "absorb" further portions of phosphorus and sulfur oxidation products.

At the end of the oxidation period, the chemical composition of slag is similar, in terms of its main components, to the chemical composition of slag after charge meltdown. One can only notice substantially higher contents of P_2O_5 and S^{2-}. While the P_2O_5 content in slag after the meltdown period is approximately from 0.1% to 0.2%, after the oxidation period it increases from 0.4% to 0.8%. Similar content ranges are obtained for sulfur. These are small amounts indeed, but the increase is by a few times.

The most important functions of the slag during the oxidation period include the dissolution of products of reactions of oxidation of the metal bath constituents, and the obtaining of properties favorable to the foaming process. Then, the slag features an oxidizing nature, a basicity between 3.0 and 3.5, and a medium viscosity. The objective of the optimization of technological conditions of slag formation during the oxidation period is to use good quality lime with an appropriate grain size (preferably from 1 to 10 mm), and an effective method of introduction, preferably by lance injection.

After the oxidation period, the slag function completely changes. Its oxidizing properties must be replaced with refining properties. In this process, there is no oxidation and the adequate metallurgical cleanness of steel is obtained by removing nonmetallic inclusions. Slag in the furnace after the oxidation period must be deoxidized already at the beginning of refining, or a new, nonoxidizing slag should be formed. Finely ground ferrosilicon or a carbon material is used to deoxidize the slag. The objective of slag deoxidation is to reduce iron oxides to a content below 2%–3%. New slag is made by adding a mixture of lime and bauxite to the furnace. An adequate viscosity and surface tension are important properties of slag in this period, and they determine good capabilities for the assimilation of nonmetallic inclusions flowing out of the metal bath.

The most important functions of slag in the refining period include the assimilation of nonmetallic inclusions forming during the metal bath deoxidation. Then, the slag features a nonoxidizing nature and its basicity ranges from 3% to 4%, at a low viscosity and a low surface tension. The objectives of optimization of the slag forming process conditions in the refining period are to deoxidize the slag as well as to ensure its low viscosity and low surface tension. At the end of the refining period, the chemical constitution of slag is as follows:

CaO: 35%–45%,
SiO_2: 10%–15%,
FeO: 1%–3%,
MgO: 8%–12%,
Al_2O_3: 15%–25%,
MnO: 5%–8%,
P_2O_5: <0.02%.

The MgO content is related to dissolving dolomite lime, and partially the furnace wall refractory lining. Steel is tapped to the ladle under a slag like this, and the slag also flows down with the metal.

If a practice is applied without the oxidation and refining periods, the slag after meltdown remains in the furnace, and the metal bath is heated prior to tapping without intentional slag forming. Changes in the slag chemistry may result from the dissolution of oxidation products of melt constituents, and changes in the physical and chemical properties related to the increase in temperature. One of the most important slag functions during melting is to increase the efficiency of heat transfer from the arc to the metal bath, which is obtained by slag foaming. Therefore, in the final period of melting, the slag should have properties favorable to foaming. If the process without oxidation and refining is applied, a typical chemical composition of the slag before tapping is as follows:

CaO: 35%–45%,
SiO_2: 10%–20%,
FeO: 15%–25%,
MgO: 8%–12%,
MnO: 5%–8%,
MgO: 8%–12%,
P_2O_5: <0.02%.

The produced steel is tapped to a ladle under a slag like this. However, from the perspective of practice, it is important to carry out tapping so that only the metal flows down into the ladle, while the slag remains in the furnace. Specialists call this "slagless" or "with slag cutoff" tapping. The reason for the need of furnace slag "cutoff" is its chemical composition, and – which is particularly important – two high a content of iron oxides. The furnace slag is oxidizing, which excludes the possibility of carrying out refining operations in a ladle with its presence. Then, a new refining slag is formed in the ladle. This is performed by adding to the ladle, prior to tapping, appropriate slag-forming materials, which should form slag with low melting temperatures and good properties for the assimilation of nonmetallic inclusions. A material called synthetic slag ensures such properties. It contains calcium aluminates. Classic synthetic slags contain 50%–55% CaO, 40%–45% Al_2O_3, to 10% SiO_2, and under 1% FeO [19], and they are produced by fusion or sintering of slag-forming components [20]. Slags with this chemical constitution feature the most advantageous physicochemical properties from the perspective of steel refining, but they are relatively expensive. Therefore, a method of ladle slag forming, where a mixture of metallurgical lime and roasted bauxite is added to the ladle, is often used in practical conditions. Composed slag-forming mixtures often contain insufficient amounts of Al_2O_3. However, during tapping, steel needs to be deoxidized in the ladle, which is accomplished by adding aluminum to the ladle in the metallic form. Al_2O_3 is the product of the deoxidizing reaction. It dissolves in the slag, making up its chemical constitution to the required, optimal level, thereby ensuring the proper refining properties. Other materials are also used to make slag-forming mixtures based on waste materials from the metallurgical industry. A patent application, where reduced BOF slag utilization was proposed, is an example [21].

9.8.2 Slag Foaming Practice

The electrical energy is the basic source of heat necessary to carry out the melting process in the EAF. Modern practices, in particular as regards the production of plain carbon steels, included measures to intensify the scrap meltdown process in order to reduce the melting time. It is possible thanks to the application of various methods for adding chemical energy. Here, we have gas-oxygen burners, oxygen lances, post-combustion lances, introducing carburization agents, etc. Another important method of process intensification, and at the same time reduction of the electric power consumption, is increasing the furnace transformer power and improving the energy utilization ratio. It is because only a part of the electrical energy transformed to heat energy in the burning arc is used for the execution of the metallurgical process. The energy utilization coefficient largely depends on the "covering" ("holding") the thermal energy radiating from the arc. At the initial phase of meltdown, the burning arc is covered with scrap, and the whole thermal energy from the arc is effectively utilized for heating and melting. However, at the final meltdown phase and at the subsequent melting process stages, the burning arc is not covered with the scrap, and a substantial part of the generated thermal energy radiates to the furnace walls and little amount of it is directly used for the process execution.

The foamy slag practice is a method that is applied in the production to enhance the effectiveness of utilization of the maximum power of the furnace transformer and to improve the energy utilization coefficient. The basic objective of this technique is to increase the thickness of the slag layer on the metal bath surface. To avoid increasing the slag mass, the thickness of this layer is increased by slag foaming. Electric arcs burning between the electrode tips and the metal bath level are then covered with a layer of this foamy slag, which is adequately thicker. The existence of foamy slag enables the maximum power of arcs and a high power factor to be applied, which involves operation with long arcs. Such operating parameters increase the efficiency of arc thermal energy transfer, and at the same time contribute to their more quiet and stable burning. All of this creates more advantageous operating conditions for the electrode positioning automatic system and is less detrimental to the power supply grid. The efficiency of heat transfer from the arc to the metal bath depends on the slag layer thickness on it. When an electrode touches the solid charge or the liquid metal bath, the electric arc does not burn, and the energy is only emitted by resistance in the place of contact. The efficiency of a process like this is approximately 14%. When a free, exposed arc burns, the process efficiency is approximately 36%. If the arc burns fully submerged in the foamy slag, the efficiency of energy transfer reaches 93% [22].

Slag foaming during steel melting in an EAF, in industrial conditions, starts immediately after forming a liquid metal bath from a part of the molten scrap and immersing the other solid scrap in it, that is, after exposing the furnace walls. To this end, the foaming agent in the form of pulverized coal, fine coke breeze, or other materials containing carbon as their basic component is injected with the air jet into the slag. Oxygen is applied to intensify the charge meltdown process from about the fifth minute after turning the power on. These two utilities are injected with lances placed on the manipulator and are entered through the main door. Usually, three lances are attached to the manipulator: two for injecting oxygen and one for the foaming agent (carbon). Methods, where lances are for injecting oxygen, foaming agent, and lime, respectively, are also used. The method of lance arrangement during operation is shown in Figures 5.5 and 5.6.

During slag foaming, the oxygen lance should be so positioned that its end is submerged in the metal bath, and the injected oxygen can penetrate into the metal bath volume. The oxygen from this lance oxidizes the components dissolved in the metal bath, first silicon, manganese and carbon, and later iron. The coal lance is positioned so that the foaming agent jet goes to the bottom part of the slag layer. All lances are made of steel pipes with an internal diameter from 18 to 38 mm, depending on the flow rate, which in turn depends on the furnace capacity. The applied oxygen flow rates are from 900 to 1,800 m^3/h, and the coal flow rates are from 20 to 50 kg/h. The grain size of the coal applied should be from 1 to 5 mm. The consumption of oxygen for slag foaming in industrial conditions is from 1.5 to 1.5 kg/Mg of steel, whereas for carbon, it is from 3 to 10 kg/Mg of steel.

The foamy slag is formed as a result of carbon oxide bubbles appearing within its volume. The occurrence of CO bubbles in slag is related to the course of reaction of iron oxide reduction with the molecules of carbon injected:

$$C_{st} + (\mathrm{FeO}) = [\mathrm{Fe}] + \{\mathrm{CO}\} \tag{9.59}$$

The following process conditions are favorable to form foamy slag in the EAF:

- *Slag basicity*: approximately 2,
- *Slag temperature*: – approximately 1,833–1,853 K (1,560°C–1,580°C),
- *Iron oxide content in the slag*: – approximately 15%,
- *MgO content in slag*: – from 8% to 12%,
- *Carbon content in slag*; – from 0.15% to 0.30%,
- Silicon content should be as low as possible.

REFERENCES

1. Tardy P.: Demand and Arising of Ferrous Scrap: Strains and Consequence, *Archives of Metallurgy and Materials*, Vol. 53, 2008, No. 2, pp. 337–343.
2. Publicity materials of Velco.
3. Karbowniczek M.: Elektrometalurgia stali; cwiczenia, AGH course book no. 1364, Wydawnictwa AGH, Krakow, 1993 (in Polish).
4. Szwej H.: System komputerowy do optymalizacji namiarowania, *Wiadomosci Hutnicze*, Vol. 42, 1986, No. 5, pp. 86–89 (in Polish).
5. Taylor C. R.: Electric Furnace Steelmaking, The Iron and Steel Society, Book Crafters Inc., Chelsea, MI, 1985.
6. Wieczorek T., Pilarczyk M.: Classification of Steel Scrap in the EAF Process Using Image Analysis Methods, *Archives of Metallurgy and Materials*, Vol. 53, 2008, No, 2, pp. 613–617.
7. Baumert J., Picco M., Weiler M., Albart P., Nyssen P.: Automated Assessment of Scrap Quality before Loading into an EAF, *Archives of Metallurgy and Materials*, Vol. 53, 2008, No. 2, pp. 345–351.
8. Kaminski P.: Wpływ palników gazowo-tlenowych na pracę pieca łukowego, Master's thesis, AGH, 2000 (in Polish).
9. Healy G. W.: A New Look at Phosphorus Distribution, *Journal of Iron and Steel Institute*, Vol. 208, No 7, 1970, pp. 664–668.
10. Grigorian V., Belyanchikov L., Stomakhin A.: Theoretical principles of electric steelmaking, Mir Publishers, Moscow, 1983.
11. Toulouevski Y., Zinurov I.: *Innovation in Electric Arc Furnaces*, Springer, Berlin, 2010.
12. Pneumatic lime/dolomite injection system in operation at North Star Bluescope Steel, More Highlights, November 2006.
13. Chatterjee A., Lindfors N., Wester J.: *Process Metallurgy of LD Steelmaking, Ironmaking and Steelmaking*, Vol. 3, No. 1, 1976, pp. 21–24.
14. Holtzer M.: *Procesy metalurgiczne i odlewnicze stopow zelaza*, Wydawnictwo Naukowe PWN, Warszawa, 2013 (in Polish).
15. Derda W., Siwka J., Nowosielski C.: Controlling of the Nitrogen Content During EAF – Technology and Continuous Casting of Steel, *Archives of Metallurgy and Materials*, Vol. 53, 2008, No. 2, pp. 523–529.
16. Karbowniczek M., Wcislo Z.: Technologia wytapiania stali w elektrycznym piecu łukowym, *Hutnik-Wiadomosci Hutnicze*, Vol. 64, 1997, No. 7, pp. 284–287 (in Polish).
17. Turkdogan E.: Deoxidation of Steel, *Journal of Iron and Steel Institute*, Vol. 210, 1972, p. 21.
18. Mazanek T., Mamro K.: *Podstawy teoretyczne metalurgii zelaza*, Wyd. Slask, Katowice, 1969 (in Polish).
19. Dziarmagowski M., Karbowniczek M.: Optimization of the Slag Formation Conditions in the Ladle Steel Refining Process, *Acta Metallurgica Slovaca*, Vol. 5, 1999, No. 3, pp. 26–30.

20. Falkus J., Dziarmagowski M., Karbowniczek M., Starczewski J., Wcisło Z.: Low melting refining and desulphurization mixture, patent PL 179474, 1996
21. Dziarmagowski M., Falkus J., Drozdz P., Karbowniczek M., Kargul T., Karwan T., Karwan-Baczewska J., Zawada B.: Method of production of a slag-forming compound for secondary steel refining in a ladle or ladle furnace, European Patent Office, patent application, EP 2213753, 14 March 2009.
22. Karbowniczek M.: *Pienienie zuzla w procesach stalowniczych*, Uczelnane Wydawnictwa Naukowo-Dydaktyczne AGH, Krakow, 1999 (in Polish).

10 Mass and Heat Balances

In both the mass and heat balances, it is about determining the inputs and outputs of a specific process. In a correctly made balance, as regard any manufacturing process, the sum of the inputs must be equal to the sum of the outputs. To be able to prepare a balance by the numerical determination of all the values, the quantitative and qualitative determination of the impacts affecting all the components of the balance is necessary.

In mass balances of the steelmaking processes, including the EAF process, the amounts and chemical compositions of both the consumed materials and the products are determined. A balance like this can be grounds for carrying out actions to improve material consumption and to maximize yields.

In the heat balances of steelmaking processes, including the EAF process, the values of all energy sources and the amounts of the usable heat obtained and the heat losses occurred are determined. A balance like this enables the process energy efficiency to be assessed, and the methods and possibilities for its improvement to be analyzed. In particular for an electric arc furnace (EAF), the heat balance can be helpful for designing furnaces (electric parameters of the power supply system, water-cooling systems, etc.) and determining the optimum operating parameters.

10.1 MASS BALANCE

The EAF steelmaking process traditionally comprises three periods: meltdown, oxidation, and refining. Often, particularly in newer solutions, some stages of the process are transferred to the secondary metallurgy. In an extreme case, the EAF steelmaking process comes down to melting the charge down and transferring all of the other process stages to the secondary metallurgy.

As the EAF steelmaking process consists of separate periods, the mass balance needs to be made separately for each period. In the case of traditional practice comprising all the process periods, and with slag replacement after the meltdown and oxidation periods, the mass balance must be made separately for the meltdown, oxidation, and refining periods. If a practice without slag replacement is applied, a single mass balance can be made for a whole heat.

To make a complete mass balance, it is necessary to know the amounts and chemical compositions of each input material, including the amount of consumable refractories, and the amounts and chemical compositions of the obtained metal bath, slag, dust, and off-gas. In practice, making such a balance is very expensive and only possible when state-of-the-art measuring instruments and methods are applied. This particularly concerns the consideration of all dust and off-gas.

In practice, often approximate mass balances of steelmaking process or balances to determine of specific process parameter are made. An example is the creation of a balance, on the basis of which it is possible to determine the amounts and chemical

DOI: 10.1201/9781003130949-10

compositions of the dust and off-gas formed, knowing the exact amounts and chemical compositions of input materials, consumed refractories and the obtained masses of steel and slag in process.

The following materials can be used as charge materials for steelmaking in an EAF; they are considered inputs of the mass balance:

- materials bringing iron and other elements being the constituents of the steel made (primarily, steel scrap and ferroalloys, and potentially pig iron or other carbon-bearing materials, such as graphite electrode scrap and carburite; in some cases, DRI products),
- slag-forming materials (mainly lime or dolomite lime, and potentially limestone, fluorite, silica materials),
- oxygen gas for oxidizing,
- materials providing additional thermal energy (coal as a carburizing agent, injected pulverized coal, other fuels (e.g. natural gas) burned in additional burners),
- deoxidizers (aluminum, ferroalloys – often ground),
- consumable refractories.

On the output side of the mass balance of the EAF process, we can specify:

- produced steel,
- slag,
- dust and off-gas.

Subject to the steel grade produced and the practice applied, only some of the above-listed components can be present on the input side. However, the above-mentioned three components are always present on the output side.

The summary of input materials (so-called metallic charge and slag-forming materials) and their chemical compositions is the input data for the calculations of the mass balance. The above-mentioned data are used to calculate the obtained chemical composition of the charge from the materials used. In the classic charge meltdown practice, using the energy of the electric arc only, without any additional heat sources, the following elements contained in the metal bath oxidize: silicon – almost all, manganese – approximately 50%, and phosphorus – 20%. In addition, the iron contained in the charge also oxidizes; however, mathematical computing of the amount of iron oxidized is very difficult. It depends on many factors, which cannot always be mathematically described. We can assume that, during the melting period, approximately 1%–2% of the iron oxidizes. The amount of oxidized iron can be calculated more accurately on the basis of the ferrous oxide content in the slag, combined with estimating the slag amount. It is assumed that other elements in the charge do not oxidize.

If any additional heat sources are applied to intensify the charge melting process, an additional item appears in the mass balance on the input side – oxygen and products of oxidation reactions of elements contained in the charge with the gaseous oxygen, if applied. In addition, the amounts of oxidized elements, as constituents of the forming metal bath, change. The amounts of oxidized elements vary depending on the type of the heat source applied, used amount of gaseous oxygen, application time, etc.

The above-mentioned data are the basis for computing the chemical composition and mass of the formed metal bath and slag after the meltdown. The obtained results are the input data for computing the chemical composition and mass of the metal bath and slag after the oxidation period, knowing the amount of the meltdown slag removed from the furnace. Oxygen gas, or possibly sometimes iron ore, is usually used for the oxidation. During oxidation, slag can also be replaced, and sometimes even a few times. The amount of oxidized components of the metal bath and the amount of slag-forming materials applied should be provided for balance calculations. Depending on the practice applied, the demand for oxygen gas or iron ore is calculated for the oxidation period.

The chemical composition and mass of the metal bath after oxidation and the amounts and chemical compositions of the deoxidizers applied, ferroalloys making up the chemical composition of the melt, and slag-forming materials are the basis for computing the chemical composition and mass of the metal bath and slag before tapping.

In addition, in order to make a mass balance, it is necessary to know the consumption of the refractory materials used for the furnace lining and refractory lining repairs by gunning. The consumption of these materials widely varies, depending on the steel grade, practice applied, tap-to-tap time, and the quality of materials themselves.

All of the above-mentioned data and computing results are the basis for preparing the final balance summary. It consists of the amounts of all the inputs and outputs. The correctness of the balance is confirmed when the sums of the inputs and outputs are equal. When preparing the balance of a specific heat, the chemical compositions of the metal bath and slag are not calculated theoretically, but they are obtained on the basis of the samples taken and their chemical compositions.

The presented rules for preparing an EAF mass balance apply to the scrap-based process. If DRI products are applied as the charge, the mass balance will change. However, the general balance layout will be the same.

It is most convenient to review the method of preparation and analysis of the mass balance using a numerical example. Computing the mass balance is very simple from the mathematical perspective, but because it involves a lot of calculations, it is very laborious and painstaking. A mass balance of a steelmaking process in an EAF can be made using computer technologies.

10.2 EXAMPLE OF A MASS BALANCE OF AN ELECTRIC ARC FURNACE STEELMAKING PROCESS

Calculations for the steelmaking process, using the classic practice with the full oxidation period, are presented as an example of a mass balance preparation. The chemical composition of the steel grade produced is given in Table 10.1.

In the practice with the full oxidation period, the charge needs to contain 0.3%–0.5% carbon above the bottom limit of its content in the steel grade made. The silicon content should not exceed 0.5%, whereas the manganese content should not be more

TABLE 10.1
Chemical Composition of the Steel Produced (Author's Work)

Element	Content (%)	Element	Content (%)
C	max. 0.22	S	Max. 0.050
Mn	1.0–1.5	Cr	Max. 0.030
Si	0.20–0.55	Ni	Max. 0.030
P	max. 0.050	Cu	Max. 0.030

TABLE 10.2
The Assumed Chemical Composition of the Pig Iron and Scrap (Author's Work)

	Chemical Composition (%)				
Material	C	Mn	Si	P	S
Pig iron	4.20	0.80	0.90	0.15	0.04
Scrap	0.20	0.50	0.40	0.02	0.03

TABLE 10.3
The Assumed Chemical Compositions of Lime and Refractory Materials (Author's Work)

	Chemical Composition (%)										
Material	CaO	SiO_2	MnO	MgO	FeO	Fe_2O_3	Al_2O_3	Cr_2O_3	P_2O_5	H_2O	CO_2
Lime	91.0	3.0	0.5	0.7	–	0.5	1.0	–	0.1	0.2	3.0
Magnesite	1.5	5.3	–	90.0	0.2	2.4	0.4	0.2	–	–	–
Chromite-magnesite	2.0	5.0	–	60.0	–	11.4	0.0	12.0	–	–	–
Magnesite mass	4.5	8.0	–	80.0	0.2	5.7	0.8	–	–	–	0.8
High alumina materials	0.3	17.7	–	0.3	0.1	1.6	80.0	–	–	–	–

than 0.8% (see Section 9.2). It is assumed that scrap applied as the charge will have a chemical composition as per Table 10.2. Pig iron, with the chemical composition also specified in Table 10.2, will be applied as the carburizing agent.

The chemical compositions of the slag-forming materials applied and consumable refractory materials are specified in Table 10.3.

Magnesite materials are usually used for bottoms, banks, and part of walls of the furnace, the least exposed to the electric arc impact. Chromite-magnesite materials are applied in the wall areas that are most exposed to the electric arc impact. The foregoing principles concern a furnace with the classic refractory lining. For water-cooled walls, only a thin refractory magnesite layer is used. The magnesite mass used for gunning almost fully passes into the slag. When water-cooling is not applied, the roof is usually made of high-alumina materials (in older solutions, silica refractories were used).

All calculations are made per 100 kg metallic charge. It is assumed that the charge consists of 90% scrap and 10% pig iron. The resulting chemical composition of the metallic charge is specified in Table 10.4.

The balance to 100 kg, that is 98.356 kg, is iron. It is assumed that, during the charge melting, no additional heat sources are applied and, therefore, carbon and sulfur will not oxidize; manganese will oxidize in 50%, silicon in whole, and phosphorus in 20%. In addition, it is assumed for the calculations [1] that, during melting, the oxidation of iron contained in the charge accounts for 2% of the charge mass. Out of this amount, 25% oxidizes to FeO, and 5% to Fe_2O_3, and both these oxides pass into the slag. The balance of oxidized iron evaporates in the arc zone. The amounts of oxidized elements during the melting period are shown in Table 10.5.

The metal bath after meltdown will have the chemical composition specified in Table 10.6.

TABLE 10.4
Chemical Composition of the Charge (Author's Work)

Element	C	Mn	Si	P	S
kg/100 kg Metallic charge	0.60	0.53	0.45	0.033	0.031

TABLE 10.5
Amounts of Oxidized Elements during the Melting Period (Author's Work)

Element	C	Mn	Si	P	S	Fe
kg/100 kg Metallic charge	0.00	0.265	0.45	0.0066	0.0	1.967

TABLE 10.6
Chemical Composition of the Metal Bath after Meltdown (Author's Work)

Element	C	Mn	Si	P	S	Fe
kg/100 kg Metallic charge	0.60	0.265	0.0	0.0264	0.0031	96.389

Chemical reactions for oxidation of the metal bath constituents are as follows:

$$[Si] + [O] = (SiO_2) \tag{10.1}$$

$$[Mn] + [O] = (MnO) \tag{10.2}$$

$$2[P] + 5[O] = (P_2O_5) \tag{10.3}$$

$$[Fe] + [O] = (FeO) \tag{10.4}$$

$$2[Fe] + 3[O] = (Fe_2O_3) \tag{10.5}$$

Therefore, we obtain:

$0.45 \cdot 60/28 = 0.964$ kg SiO_2/100 kg metallic charge,
$0.265 \cdot 71/55 = 0.342$ kg MnO/100 kg metallic charge,
$0.0066 \cdot 142/62 = 0.0151$ kg P_2O_5/100 kg metallic charge,
$1.967 \cdot 72/56 \cdot 0.25 = 0.632$ kg FeO/100 kg metallic charge,
$1.967 \cdot 160/112 \cdot 0.05 = 0.141$ kg Fe_2O_3/100 kg metallic charge.

The oxygen demand for oxidizing the above-mentioned constituents is as follows:

- to oxidize silicon to SiO_2:

$$0.45 \cdot 32/28 = 0.514 \text{ kg oxygen/100 kg metallic charge,}$$

- to oxidize manganese to MnO:

$$0.265 \cdot 16/55 = 0.077 \text{ kg oxygen/100 kg metallic charge,}$$

- to oxidize phosphorus to P_2O_5:

$$0.0066 \cdot 80/62 = 0.0085 \text{ kg oxygen/100 kg metallic charge,}$$

- to oxidize iron to FeO:

$$1.967 \cdot 16/56 \cdot 0.25 = 0.141 \text{ kg oxygen/100 kg metallic charge,}$$

- to oxidize iron to Fe_2O_3:

$$1.967 \cdot 48/112 \cdot 0.05 = 0.042 \text{ kg oxygen / 100 kg metallic charge.}$$

The total oxygen demand in the melting period is 0.7825 kg/100 kg metallic charge.

It is assumed that, in order to create slag, 3.5 kg lime is added per 100 kg metallic charge. During gunning, 0.1 kg magnesite mass is used per 100 kg metallic charge, and the total of magnesite mass passes into the slag during melting. The consumption of refractory materials is as follows:

- magnesite materials: 0.7 kg/100 kg metallic charge,
- chromite-magnesite materials: 0.9 kg/100 kg metallic charge,
- high-alumina materials: 0.6 kg/100 kg metallic charge.

In addition, it is assumed that the distribution of consumption of refractory materials and the gunning magnesite mass during melting is as follows:

- 60% is consumed during melting,
- 30% is consumed during oxidation,
- 10% is consumed during refining.

With these assumptions, the amounts of individual compounds of the forming slag during the melting period will be as follows:

$$CaO\text{: } 3.5\cdot 0.91 + 0.7\cdot 0.60\cdot 0.015 + 0.9\cdot 0.60\cdot 0.02 + 0.1\cdot 0.60\cdot 0.045 + 0.60\cdot 0.60\cdot 0.003 = 3.206 \text{ kg/100 kg metallic charge,}$$

$$SiO_2\text{: } 3.5\cdot 0.003 + 0.7\cdot 0.60\cdot 0.053 + 0.9\cdot 0.60\cdot 0.056 + 0.1\cdot 0.60\cdot 0.08 + 0.60\cdot 0.60\cdot 0.177 + 0.964 = 1.190 \text{ kg/100 kg metallic charge,}$$

$$MnO\text{: } 3.5\cdot 0.005 + 0.342 = 0.360 \text{ kg/100 kg metallic charge,}$$

$$MgO\text{: } 3.5\cdot 0.007 + 0.7\cdot 0.60\cdot 0.90 + 0.9\cdot 0.60\cdot 0.60 + 0.1\cdot 0.60\cdot 0.80 + 0.60\cdot 0.60\cdot 0.003 = 0.776 \text{ kg/100 kg metallic charge,}$$

$$FeO\text{: } 0.7\cdot 0.60\cdot 0.002 + 0.1\cdot 0.60\cdot 0.002 + 0.60\cdot 0.60\cdot 0.001 + 0.632 = 0.634 \text{ kg/100 kg metallic charge,}$$

$$Fe_2O_3\text{: } 3.5\cdot 0.005 + 0.7\cdot 0.60\cdot 0.0024 + 0.9\cdot 0.60\cdot 0.114 + 0.1\cdot 0.60\cdot 0.057 + 0.60\cdot 0.60\cdot 0.016 + 0.141 = 0.239 \text{ kg/100 kg metallic charge,}$$

$$Al_2O_3\text{: } 3.5\cdot 0.01 + 0.7\cdot 0.60\cdot 0.004 + 0.9\cdot 0.60\cdot 0.09 + 0.1\cdot 0.60\cdot 0.008 + 0.60\cdot 0.60\cdot 0.80 = 0.374 \text{ kg/100 kg metallic charge,}$$

$$Cr_2O_3\text{: } 0.7 \cdot 0.60 \cdot 0.002 + 0.9 \cdot 0.60 \cdot 0.12 = 0.066 \text{ kg/100 kg metallic charge,}$$

$$P_2O_5\text{: } 3.5 \cdot 0.001 + 0.0151 = 0.0186 \text{ kg/100 kg metallic charge.}$$

The final amount and chemical composition of the meltdown period slag is shown in Table 10.7.

In the classic practice with the full oxidizing period, this period starts with removing some of the meltdown slag and adding an appropriate amount of lime to form a new one. Oxidation is performed with an oxygen lance. During the oxidation period, the carbon content in the metal bath decreases by 0.45 percentage points. At the same time, manganese is oxidized in the amount of 40%, phosphorus in the amount of 40%, and sulfur in the amount of 50%, in relation to their initial content. In addition, it is assumed for the calculations that, during oxidation, the iron contained in the metal bath oxidizes in 0.5% of the melt mass. As in the melting period, 25% oxidizes to FeO, and 5% to Fe_2O_3, which pass into the slag, and the balance evaporates. The amounts of oxidized components of the metal bath during the oxidation period are shown in Table 10.8.

The chemical composition of the metal bath after oxidation is shown in Table 10.9.

TABLE 10.7
Final Amount and Chemical Composition of the Meltdown Period Slag (Author's Work)

	Chemical Composition									
Compound	**CaO**	**SiO_2**	**MnO**	**MgO**	**FeO**	**Fe_2O_3**	**Al_2O_3**	**Cr_2O_3**	**P_2O_5**	**Total**
kg/100 kg Metallic charge	3.206	1.190	0.360	0.776	0.634	0.239	0.374	0.066	0.019	6.862
%	46.72	17.35	5.24	11.30	9.23	3.48	5.45	0.96	0.27	100

TABLE 10.8
Amounts of Oxidized Elements during the Oxidation Period (Author's Work)

Element	C	Mn	Si	P	S	Fe
kg/100 kg Metallic charge	0.45	0.106	0.00	0.0106	0.0155	0.482

TABLE 10.9
Chemical Composition of the Metal Bath after Oxidation (Author's Work)

Element	C	Mn	Si	P	S	Fe
kg/100 kg Metallic charge	0.15	0.159	0.00	0.0158	0.0155	95.907

The chemical reactions for oxidation of the metal bath constituents are analogous to the ones described by equations (10.1)–(10.5). However, carbon oxidizes as per the reaction:

$$[C]+[O]=\{CO\} \tag{10.6}$$

while sulfur passes from the metal bath to the slag as per the reaction:

$$[S]+\left(O^{2-}\right)=\left(S^{2-}\right)+[O] \tag{10.7}$$

Therefore, during oxidation we obtain:

$0.45 \cdot 28/12 = 1.050$ kg CO/100 kg metallic charge,
$0.106 \cdot 71/55 = 0.137$ kg MnO/100 kg metallic charge,
$0.0106 \cdot 142/62 = 0.0243$ kg P_2O_5/100 kg metallic charge,
$0.482 \cdot 72/56 \cdot 0.25 = 0.155$ kg FeO/100 kg metallic charge,
$0.482 \cdot 160/112 \cdot 0.05 = 0.034$ kg Fe_2O_3/100 kg metallic charge.

The oxygen demand for oxidizing the above-mentioned constituents is as follows:

- to oxidize carbon to CO:

$$0.45 \cdot 16/12 = 0.600 \text{ kg oxygen}/100 \text{ kg metallic charge},$$

- to oxidize manganese to MnO:

$$0.106 \cdot 16/55 = 0.031 \text{ kg oxygen}/100 \text{ kg metallic charge},$$

- to oxidize phosphorus to P_2O_5:

$$0.0106 \cdot 80/62 = 0.0137 \text{ kg oxygen}/100 \text{ kg metallic charge},$$

- to oxidize iron to FeO:

$$0.482 \cdot 16/56 \cdot 0.25 = 0.034 \text{ kg oxygen}/100 \text{ kg metallic charge},$$

- to oxidize iron to Fe_2O_3:

$$0.482 \cdot 48/112 \cdot 0.05 = 0.010 \text{ kg oxygen}/100 \text{ kg metallic charge}.$$

The total oxygen demand in the oxidation period is 0.6887 kg/100 kg metallic charge.

During the oxidation period, the assumed 0.0155 kg sulfur should be removed from the metal bath to the slag. To make it possible, the basicity of the oxidation period slag should be at least 3.0 (in this example, the assumed slag basicity is 3.5). To this end, some part of the oxidation slag should be removed and an adequate amount of lime should be added. These amounts can be calculated from the relationship:

$$\frac{CaO}{SiO_2} = B \tag{10.8}$$

where:

CaO: $3.206 \cdot y/100 + x \cdot 91/100 + 0.7 \cdot 30/100 \cdot 1.5/100 + 0.9 \cdot 30/100 \cdot 2.0/100 + 0.1 \cdot 30/100 \cdot 4.5/100 + 0.60 \cdot 30/100 \cdot 0.3/100 = 0.3206 \cdot y + 0.91 \cdot x + 0.1044$

SiO_2: $1.190 \cdot y/100 + x \cdot 3/100 + 0.7 \cdot 30/100 \cdot 5.3/100 + 0.9 \cdot 30/100 \cdot 5.6/100 + 0.1 \cdot 30/100 \cdot 8.0/100 + 0.60 \cdot 30/100 \cdot 17.7/100 = 0.0119 \cdot y + 0.03 \cdot x + 0.06051$

Where B means the slag basicity, y the % share of the slag remaining in the furnace, and x the amount of added lime in kg/100 kg metallic charge. With the above-mentioned assumptions, the equation (10.8) will have the form:

$$\frac{0.3206y + 0.91x + 0.1044}{0.0119y + 0.03x + 0.06051} = 3.5 \tag{10.9}$$

Therefore, after transformations:

$$x = 0.0119 \cdot y + 0.250 \tag{10.10}$$

For instance, if we want to remove 30% meltdown period slag, we need to add 1.085 kg lime to obtain the basicity of the oxidation period slag of 3.5. For further calculations, it is assumed that 50% meltdown slag is removed. Then, we need to add 0.847 kg lime. With these assumptions, the amounts of individual compounds of the forming slag during the oxidation period will be as follows:

CaO: $3.206 \cdot 50/100 + 0.847 \cdot 91/100 + 0.7 \cdot 0.30 \cdot 1.5/100 + 0.9 \cdot 0.30 \cdot 2.0/100 + 0.1 \cdot 0.30 \cdot 4.5/100 + 0.60 \cdot 0.30 \cdot 0.3/100 = 2.384$ kg/100 kg metallic charge,

SiO_2: $1.190 \cdot 50/100 + 0.847 \cdot 3/100 + 0.7 \cdot 0.30 \cdot 5.3/100 + 0.9 \cdot 0.30 \cdot 5.6/100 + 0.1 \cdot 0.30 \cdot 8.0/100 + 0.60 \cdot 0.30 \cdot 17.7/100 = 0.681$ kg/100 kg metallic charge,

MnO: $0.360 \cdot 50/100 + 0.847 \cdot 1.5/100 + 0.137 = 0.321$ kg/100 kg metallic charge,

MgO: $0.776 \cdot 50/100 + 0.847 \cdot 0.7/100 + 0.7 \cdot 0.30 \cdot 90/100 + 0.9 \cdot 0.30 \cdot 60/100 + 0.1 \cdot 0.30 \cdot 80/100 + 0.60 \cdot 0.30 \cdot 0.3/100 = 0.769$ kg/100 kg metallic charge,

FeO: $0.634 \cdot 50/100 + 0.7 \cdot 0.30 \cdot 0.2/100 + 0.1 \cdot 0.30 \cdot 0.2/100 + 0.60 \cdot 0.30 \cdot 0.1/100 + 0.155 = 0.472$ kg/100 kg metallic charge,

Fe_2O_3: $0.239 \cdot 50/100 + 0.847 \cdot 0.5/100 + 0.7 \cdot 0.30 \cdot 2.4/100 + 0.9 \cdot 0.30 \cdot 11.4/100 + 0.1 \cdot 0.30 \cdot 5.7/100 + 0.60 \cdot 0.30 \cdot 1.6/100 + 0.034 = 0.198$ kg/100 kg metallic charge,

Al_2O_3: $0.374 \cdot 50/100 + 0.847 \cdot 1.0/100 + 0.7 \cdot 0.30 \cdot 0.4/100 + 0.9 \cdot 0.30 \cdot 9.0/100 + 0.1 \cdot 0.30 \cdot 0.8/100 + 0.60 \cdot 0.30 \cdot 80/100 = 0.365$ kg/100 kg metallic charge,

Cr_2O_3: $0.066 \cdot 50/100 + 0.7 \cdot 0.30 \cdot 0.2/100 + 0.9 \cdot 0.30 \cdot 12/100 = 0.066$ kg/100 kg metallic charge,

P_2O_5: $0.019 \cdot 50/100 + 0.847 \cdot 0.3/100 + 0.0243 = 0.035$ kg/100 kg metallic charge.

The final amount and chemical composition of the oxidation period slag is shown in Table 10.10. Sulfur that passes into the slag according to reaction (10.7) is not considered a slag constituent, whereas its amount is added to the slag mass; therefore, the sum of slag constituents in Table 10.10 is lower than the value presented in the column "total", exactly by the amount of sulfur arising from Table 10.9.

After the completion of the oxidation period, the metal bath is initially deoxidized. In this example, it is assumed that 0.34 kg technical aluminum was used per 100 kg metallic charge for initial deoxidizing. For further deoxidizing and to make up silicon and manganese in the metal bath, 0.25 kg FeSi75 and 0.5 kg FeSiMn were added per 100 kg metallic charge. The chemical composition of the ferroalloys applied is given in Table 10.11.

After adding the ferroalloys, the chemical composition of the metal bath will change, and the contents of individual elements will be as follows:

C: $0.15 + 0.25 \cdot 0.15/100 + 0.5 \cdot 1.5/100 = 0.158$ kg/100 kg metallic charge,
Mn: $0.159 + 0.25 \cdot 0.5/100 + 0.5 \cdot 65/100 = 0.485$ kg/100 kg metallic charge,
Si: $0.25 \cdot 75/100 + 0.5 \cdot 18/100 = 0.278$ kg/100 kg metallic charge,
P: $0.0158 + 0.25 \cdot 0.06/100 + 0.5 \cdot 0.1/100 = 0.0165$ kg/100 kg metallic charge,
S: $0.0155 + 0.25 \cdot 0.04/100 + 0.5 \cdot 0.02/100 = 0.0157$ kg/100 kg metallic charge,
Al: $0.34 \cdot 97/100 = 0.330$ kg/100 kg metallic charge,
Fe: $95.907 + 0.34 \cdot 3/100 + 0.25 \cdot 24.25/100 + 0.5 \cdot 15.38/100 = 96.055$ kg/100 kg metallic charge.

TABLE 10.10
Final Amount and Chemical Composition of the Oxidation Period Slag (Author's Work)

	Chemical Composition									
Compound	**CaO**	**SiO_2**	**MnO**	**MgO**	**FeO**	**Fe_2O_3**	**Al_2O_3**	**Cr_2O_3**	**P_2O_5**	**Total**
kg/100 kg Metallic charge	2.384	0.681	0.321	0.769	0.472	0.198	0.365	0.066	0.035	5.306
%	44.93	12.84	6.05	14.50	8.90	3.74	6.87	1.24	0.66	100

TABLE 10.11
Chemical Composition of Ferroalloys Used in the Calculations (Author's Work)

	Chemical Composition (%)						
Material	**C**	**Mn**	**Si**	**P**	**S**	**Al**	**Fe**
FeSi75	0.15	0.5	75.0	0.06	0.04	–	24.25
FeMn	7.0	65.0	–	0.35	–	–	27.65
FeSiMn	1.5	65.0	18.0	0.1	0.02	–	15.38
Technical Al	–	–	–	–	–	97.0	3.0

From the industrial practice data, it follows that, after adding ferroalloys to the initial deoxidation, the following elements oxidize: 95% aluminum, 70% silicon, and 40% manganese. Therefore, during the initial deoxidation, the following will oxidize:

$0.485 \cdot 40/100 = 0.194$ kg Mn/100 kg metallic charge,
$0.278 \cdot 70/100 = 0.195$ kg Si/100 kg metallic charge,
$0.330 \cdot 95/100 = 0.314$ kg Al/100 kg metallic charge.

Manganese and silicon oxidize according to reactions (10.2) and (10.1), respectively. Aluminum oxidizes according to the reaction:

$$2[\mathrm{Al}] + 3[\mathrm{O}] = (\mathrm{Al_2O_3}) \qquad (10.11)$$

At the same time, the following amount of oxides will form and pass to the slag:

$0.194 \cdot 71/55 = 0.250$ MnO/100 kg metallic charge,
$0.195 \cdot 60/28 = 0.418$ SiO_2/100 kg metallic charge,
$0.314 \cdot 102/54 = 0.593$ Al_2O_3/100 kg metallic charge.

The oxygen demand from the metal bath to oxidize the above-mentioned elements is as follows:

- to oxidize manganese to MnO:

  $0.194 \cdot 16/55 = 0.0564$ kg oxygen/100 kg metallic charge,

- to oxidize silicon to SiO_2:

  $0.195 \cdot 32/28 = 0.2229$ kg oxygen/100 kg metallic charge,

- to oxidize aluminum to Al_2O_3:

  $0.314 \cdot 48/54 = 0.2787$ kg oxygen/100 kg metallic charge.

The total oxygen demand during the initial deoxidation is 0.558 kg/100 kg metallic charge.

After the initial deoxidation, the content of oxidizing elements in the metal bath will be as follows:

$0.485 \cdot 60/100 = 0.291$ kg Mn/100 kg metallic charge,
$0.278 \cdot 30/100 = 0.083$ kg Si/100 kg metallic charge,
$0.330 \cdot 5/100 = 0.017$ kg Al/100 kg metallic charge.

The ultimate chemical composition of the metal bath after initial deoxidization is shown in Table 10.12.

TABLE 10.12
The Ultimate Chemical Composition of the Metal Bath after Initial Deoxidization (Author's Work)

Element	C	Mn	Si	P	S	Al	Fe
kg/100 kg Metallic charge	0.158	0.291	0.083	0.0165	0.0157	0.017	96.055

Taking into account the assumed chemical composition of steel to be produced (Table 10.1), to make up the chemical composition of the metal bath, 0.709–1.209 kg manganese and 0.117–0.467 kg silicon should be added per 100 kg metallic charge. It is known from industrial practice that the "melting loss" of deoxidizers added after the initial deoxidation is approximately 5% manganese and 45% silicon, respectively. Therefore, ferroalloy amounts that should be added to make up the chemical composition can be calculated as follows:

- we assume that the metal bath should be made up with 0.8 kg manganese and 0.2 kg silicon
- considering the assumed oxidation percentage, one need to add:

$0.8 \cdot 100/95 = 0.842$ kg manganese/100 kg metallic charge,
$0.2 \cdot 100/55 = 0.364$ kg silicon/100 kg metallic charge.

- we assume that the missing amounts of manganese and silicon are added using the following ferroalloys: FeMn (the cheapest one, but it contains a lot of carbon), FeSiMn, and FeSi75
- taking into account that the manganese content in FeMn and in FeSiMn is the same (65%), regardless of the used type of ferroalloy, one should add:

$0.842 \cdot 100/65 = 1.295$ kg FeMn or FeSiMn/100 kg metallic charge,

- the maximum addition of FeMn should be calculated not to exceed the carbon limit in the metal bath.

FeMn contains 7% carbon. Before adding the ferroalloys, there is 0.208 kg carbon in the metal bath. Therefore, the amount of carbon brought by FeMn must not exceed 0.03 kg, as after taking into account all other components bringing carbon into the metal bath the carbon content must not exceed 0.22%. Therefore, the acceptable amount of FeMn can be at most: $0.03 \cdot 100/7 = 0.429$ kg FeMn/100 kg metallic charge. To ensure an appropriate amount of manganese in the melt, one should add $1.295 - 0.429 = 0.866$ kg FeSiMn,

- FeSiMn brings $0.866 \cdot 18/100 = 0.156$ kg silicon per 100 kg metallic charge,
- 0.364 kg silicon should be added to the melt and, therefore, the balance or $0.364 - 0.156 = 0.208$ kg should be made up with FeSi75,

- the amount of FeSi75 added should be:

$$0.208 \cdot 100/75 = 0.277 \text{ kgFeSi75/100 kg metallic charge.}$$

To conclude, the following should be added to finally deoxidize the metal bath and to make up its chemical composition:

- 0.429 kg FeMn,
- 0.866 kg FeSiMn,
- 0.277 kg FeSi75.

Adding the above-mentioned amounts of ferroalloys will bring:

C: $0.429 \cdot 7/100 + 0.866 \cdot 1.5/100 + 0.277 \cdot 0.15/100 = 0.043$ kg/100 kg metallic charge,
Mn: $0.429 \cdot 65/100 + 0.866 \cdot 65/100 + 0.277 \cdot 0.5/100 = 0.843$ kg/100 kg metallic charge,
Si: $0.866 \cdot 18/100 + 0.277 \cdot 75/100 = 0.364$ kg/100 kg metallic charge,
P: $0.429 \cdot 0.35/100 + 0.866 \cdot 0.1/100 + 0.277 \cdot 0.06/100 = 0.0025$ kg/100 kg metallic charge,
S: $0.866 \cdot 0.02/100 + 0.277 \cdot 0.04/100 = 0.0003$ kg/100 kg metallic charge,
Fe: $0.429 \cdot 27.65/100 + 0.866 \cdot 15.38/100 + 0.277 \cdot 24.25/100 = 0.319$ kg/100 kg metallic charge.

Furthermore, out of the elements brought, only 0.8 kg manganese and 0.2 kg silicon will pass into the metal bath; the rest will oxidize and pass into the slag.

At the last stage of melting before tapping, according to the practice applied, favorable conditions for the return reaction of phosphorus occur in the EAF. From the industrial practice data, it follows that the resulting increase in the phosphorus content in the metal bath is in the range of 10–40%. Assuming that the phosphorus content increase in the example concerned is 20%, its content in the finished steel will be:

$$(0.0165 + 0.0025) \cdot 120/100 = 0.0228 \text{ kg/100 kg metallic charge.}$$

Therefore, $0.0228 - (0.0165 + 0.0025) = 0.0038$ kg phosphorus will leave the slag. This means a loss of:

$$0.0038 \cdot 142/62 = 0.0087 \text{ kg } P_2O_5\text{/100 kg metallic charge.}$$

At the same time, the following will be emitted:

$$0.0038 \cdot 89/62 = 0.0049 \text{ kg oxygen/100 kg metallic charge.}$$

The ultimate chemical composition of the metal bath before tapping is shown in Table 10.13.

TABLE 10.13
Chemical Composition of the Metal Bath before Tapping (Author's Work)

Element	C	Mn	Si	P	S	Al	Fe	Total
kg/100 kg Metallic charge	0.201	1.091	0.283	0.023	0.016	0.017	96.374	98.005
%	0.205	1.113	0.289	0.023	0.016	0.017	98.337	100

Taking into account the assumed melting loss of ferroalloys added to final deoxidation and to make up the chemical composition of the metal bath, the following will oxidize:

$0.842 - 0.8 = 0.042$ kg manganese/100 kg metallic charge,
$0.364 - 0.2 = 0.164$ kg silicon/100 kg metallic charge.

At the same time, the following will form and pass to the slag (as per reactions (10.2) and (10.1)):

$0.042 \cdot 71/55 = 0.054$ kg MnO/100 kg metallic charge,
$0.164 \cdot 60/28 = 0.351$ kg SiO_2/100 kg metallic charge.

The oxygen demand for oxidizing the above-mentioned elements is as follows:

- to oxidize manganese to MnO:

$$0.042 \cdot 16/55 = 0.0122 \text{ kg oxygen/100 kg metallic charge,}$$

- to oxidize silicon to SiO_2:

$0.164 \cdot 32/28 = 0.1874$ kg oxygen/100 kg metallic charge.

Taking into account 10% consumption of the refractory materials in this period of melting in relation to their total consumption, the amounts of individual compounds of the forming slag in the refining period are as follows:

CaO: $2.384 + 0.7 \cdot 10/100 \cdot 1.5/100 + 0.9 \cdot 10/100 \cdot 2.0/100 + 0.1 \cdot 10/100 \cdot 4.5/100 + 0.60 \cdot 10/100 \cdot 0.3/100 = 2.387$ kg/100 kg metallic charge,
SiO_2: $0.681 + 0.7 \cdot 10/100 \cdot 5.3/100 + 0.9 \cdot 10/100 \cdot 5.6/100 + 0.1 \cdot 10/100 \cdot 8.0/100 + 0.60 \cdot 10/100 \cdot 17.7/100 + 0.418 + 0.351 = 1.461$ kg/100 kg metallic charge,
MnO: $0.321 + 0.250 + 0.054 = 0.625$ kg/100 kg metallic charge,
MgO: $0.769 + 0.7 \cdot 10/100 \cdot 90/100 + 0.9 \cdot 10/100 \cdot 60/100 + 0.1 \cdot 10/100 \cdot 80/100 + 0.60 \cdot 10/100 \cdot 0.3/100 = 0.894$ kg/100 kg metallic charge,
FeO: $0.472 + 0.7 \cdot 10/100 \cdot 0.2/100 + 0.1 \cdot 10/100 \cdot 0.2/100 + 0.60 \cdot 10/100 \cdot 0.1/100 = 0.472$ kg/100 kg metallic charge,
Fe_2O_3: $0.198 + 0.7 \cdot 10/100 \cdot 2.4/100 + 0.9 \cdot 10/100 \cdot 11.4/100 + 0.1 \cdot 10/100 \cdot 5.7/100 + 0.60 \cdot 10/100 \cdot 1.6/100 = 0.211$ kg/100 kg metallic charge,

Al_2O_3: $0.365 + 0.7 \cdot 10/100 \cdot 0.4/100 + 0.9 \cdot 10/100 \cdot 9.0/100 + 0.1 \cdot 10/100 \cdot 0.8/100 + 0.60 \cdot 10/100 \cdot 80/100 + 0.593 = 1.014$ kg/100 kg metallic charge,

Cr_2O_3: $0.066 + 0.7 \cdot 10/100 \cdot 0.2/100 + 0.9 \cdot 10/100 \cdot 12/100 = 0.077$ kg/100 kg metallic charge,

P_2O_5: $0.035 - 0.0087 = 0.0263$ kg/100 kg metallic charge.

The ultimate chemical composition of the slag before tapping is shown in Table 10.14.

In this case, as after the oxidation, the mass of slag includes the sulfur content although it is not shown in a separate column.

All the above-mentioned calculations are the basis for the final summary of mass balance of a heat. A summary like this is shown in Table 10.15. In addition, the amounts of gases occurring on both the input and output sides are the sum of respective gases from subsequent periods of the melting. On the input side, there is atmospheric oxygen used to oxidize constituents of the metal bath; the amount of oxygen

TABLE 10.14
Amount and Chemical Composition of the Slag before Tapping (Author's Work)

	Chemical Composition									
Compound	CaO	SiO_2	MnO	MgO	FeO	Fe_2O_3	Al_2O_3	Cr_2O_3	P_2O_5	Total
kg/100 kg Metallic charge	2.387	1.461	0.625	0.894	0.472	0.211	1.014	0.077	0.026	7.183
%	44.93	12.84	6.05	14.50	W	3.74	6.87	1.24	0.66	100

TABLE 10.15
Mass Balance of the Electric Arc Furnace Process (Author's Work)

Introduced Material	kg	%	Obtained Material	kg	%
1. Scrap	90.00	80.70	1. Steel	98.005	87.87
2. Pig iron	10.00	8.97	2. Slag	10.614	9.52
3. Lime	4.34	3.89	3. CO	1.050	0.94
Refractories			4. CO_2	0.131	0.12
4. Magnesite materials	0.7	0.63	5. H_2O	0.009	0.01
5. Chromite-magnesite	0.9	0.81	6. Ignition loss	1.714	1.54
6. High alumina	0.6	0.54	Inaccuracies	0.003	0.00
7. Magnesite mass	0.1	0.09			
8. Atmospheric oxygen	2.224	1.99			
9. Technical aluminum	0.34	0.31			
10. FeSi75	0.527	0.47			
11. FeSiMn	1.366	1.22			
12. FeMn	0.429	0.38			
Total	111.52	100	Total	111.52	100

emitted during the return reduction of phosphorus is also included. On the output side, there is CO emitted during the oxidation of the carbon contained in the metal bath as well as the CO_2 and H_2O contained in the lime and magnesite mass. In this case, as after the oxidation, the mass of slag includes the sulfur content although it is not shown in a separate column.

10.3 HEAT BALANCE

The heat balance of the EAF steelmaking process can be determined for the whole melting process, from tap to tap, as well as in certain, rare cases, for its selected period or periods [2,3]. A prepared heat balance is the foundation for the furnace thermal operation analysis.

For the EAF process, preparing a heat balance is less important for the practice than for the BOF process. In the BOF process, the thermal balance is the basis for the correct composition of the charge (quantitative selection of hot metal and scrap shares, taking into account their chemical compositions), and the related steelmaking management. In this process, only exothermic chemical reactions are the source of heat, and if the process is off-balance, we can obtain heat with an incorrect chemical composition or incorrect temperature. This problem does not occur in an EAF, as its main heat source is the electrical energy converted into heat energy in the arc. Any amount of energy can be introduced to the furnace. It is determined by obtaining an appropriate temperature of liquid steel, and business considerations. As the electrical energy is expensive, the primary goal in the heat production process is to minimize its consumption. The heat balance of the EAF process can be used for this purpose.

In order to determine the full heat balance of the EAF process, it is necessary to know the volumes of all heat sources and the quantity of heat accumulated in the liquid steel, slag, and all heat losses. In practice, preparing the full heat balance is very difficult primarily because the values of all the components of heat losses cannot be accurately determined. The primary components of the EAF process heat balance are presented in Figure 10.1.

The process heat sources are:

- electrical energy, with its amount depending on the furnace design, practice applied, grades of steel made, etc.; it constitutes 80%–90% of the total thermal energy on the input side for the classic practice without additional heat sources [3]
- heat from exothermic reactions (depending on the practice applied and the steel grades manufactured, various exothermic reactions occur, e.g. the oxidation of carbon, silicon, manganese, phosphorus, and iron; in the classic practice, the amount of heat from this origin reaches a few percent, and if oxygen lances are applied, the amount of heat from exothermic reactions can reach up to 30%, or more) [4,5],
- heat from the combustion of fuels in burners; natural gas is the most common fuel; in the case of gas burners, the amount of heat from combustion can reach up to 20% [4, 5],

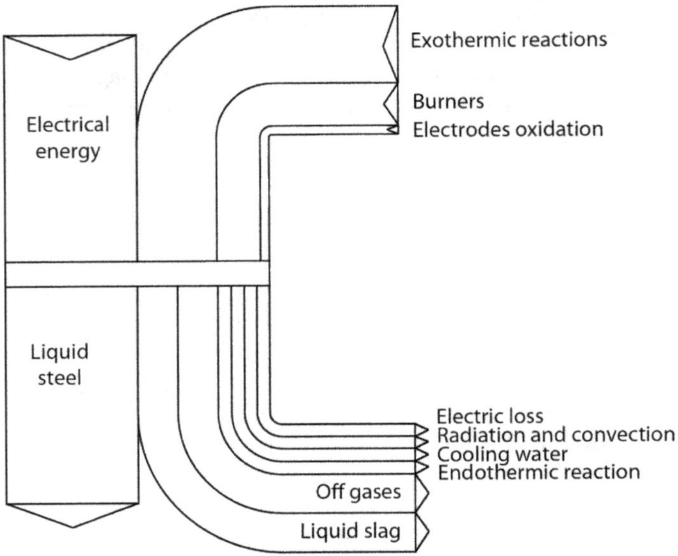

FIGURE 10.1 General diagram of the heat balance of the electric arc furnace steelmaking process. (Author's work.)

- heat from exothermic reactions occurring during the oxidation of graphite electrodes; it is assumed that approximately 40%–50% of total electrode consumption accounts for the oxidation (carbon from electrodes reacts with atmospheric oxygen producing CO) – the resulting amount of heat depends on the electrode consumption and usually does not exceed 4%,
- physical heat of charge materials (for a practice without preheating charge materials, their physical heat equals zero; if preheated scrap or preheated ferroalloys are used, their physical heat can reach a few percent in the heat balance).

On the output side of the heat balance, we can distinguish [3–5]:

- physical heat of liquid steel; it accounts for 50%–65% output heat in the balance,
- physical heat of slag; it accounts for 8%–12% output heat in the balance,
- physical heat of gases; depending on the practice applied, it accounts for 10%–20% output heat in the balance,
- endothermic reaction heat; often neglected in balance calculations due to a very low value,
- heat loss with cooling water; depending on the scope of application of water-cooling, it accounts for 5%–15% output heat in the balance,
- heat loss by radiation and convection from the outer surface of the furnace shell and roof, the inner surface of the roof, walls, and bottom during charge loading; it depends on the scope of application of water-cooling

and accounts for 5%–15% of output heat in the balance if the traditional refractory lining is applied, and under 5% if the walls and roof are cooled with water,

- electric loss related to a heat loss resulting from the current flow in a conductor; it comprises a loss in the furnace transformer, secondary circuit, and electrodes; its amount primarily depends on the applied current intensity and the design of electrical equipment; it accounts for 5%–15% output heat in the balance.

A fairly accurate estimation of the component volumes at the input side in the heat balance does not pose serious difficulties. It is more difficult to determine the volumes of components on the output side of the balance.

In the most general form, the heat balance equation can be written as follows:

$$\sum Q_p = \sum Q_r \tag{10.12}$$

where:

the left hand side means the total input heat and the right hand side means the total output heat.

Similar to the mass balance, it is the most convenient to see the method of preparation and to analyze a thermal balance using a numerical example. An example of a balance like this is presented in the following section.

10.4 EXAMPLE OF A HEAT BALANCE OF AN ELECTRIC ARC FURNACE STEELMAKING PROCESS

The mass balance is the foundation for preparing the thermal balance. The heat balance items on the input side can be calculated as follows:

1. *The amount of heat coming from the transformation of electrical energy into heat* in the electric arc is the easiest to calculate. Each furnace has electrical energy meters installed to enable the amount of electrical energy consumed to be accurately measured. It should only be converted into heat energy. This is usually the balance closing item. In the example concerned, the amount of consumed electrical energy will be calculated at the end of the balance in order to close it.
2. *The heat of additional burners* is also relatively easy to establish. Knowing the consumption of fuels used and their calorific values, one can calculate the amount of heat from this source. In the example considered, it is assumed that no additional burners are used and the amount of their heat is zero.
3. Exothermic reaction heat: The thermal effect from exothermic reactions, which can usually occur during melting in an EAF, is as follows [2]:

$$C \rightarrow CO_2\text{: } 31.81\,\text{MJ/kgC} \tag{10.13}$$

$$Fe \rightarrow Fe_3O_4: \ 5.90\,MJ/kg\,Fe \tag{10.14}$$

$$Fe \rightarrow Fe_2O_3: \ 7.35\,MJ/kg\,Fe \tag{10.15}$$

$$Fe \rightarrow FeO: \ 4.81\,MJ/kg\,Fe \tag{10.16}$$

$$Mn \rightarrow MnO: \ 7.34\,MJ/kg\,Mn \tag{10.17}$$

$$Si \rightarrow SiO_2: 28.86\,MJ/kg\,Si \tag{10.18}$$

$$P \rightarrow P_2O_5: \ 23.14\,MJ/kg\,P \tag{10.19}$$

$$Al \rightarrow Al_2O_3: \ 28.52\,MJ/kg\ Al \tag{10.20}$$

The above-mentioned data are established for the temperature of 1,873 K; it is only for reaction (10.13) that the data apply to the temperature of 293 K.

In the example concerned, the amount of heat coming from exothermic reactions is as follows:

- oxidation of carbon in the EAF process proceeds as per reaction (10.6); in practice, the emitted CO burns mostly above the metal bath to CO_2 [1]; the thermal effect from oxidizing carbon to CO_2 is therefore:

$$0.45\,kg\ C \cdot 31.81\ MJ/kg = 14.315\ MJ/100\,kg\ \text{metallic charge},$$

- oxidation of silicon:

$$(0.45 + 0.195 + 0.164)\ kg\ Si \cdot 28.86\ MJ/kg = 23.348\ MJ/100\,kg\ \text{metallic charge},$$

- oxidation of manganese:

$$(0.265 + 0.106 + 0.194 + 0.042)\ kg\ Mn \cdot 7.34\ MJ/kg = 4.455\ MJ/100\,kg\ \text{metallic charge},$$

- oxidation of phosphorus:

$$(0.0066 + 0.0106 - 0.0038)\ kg\ P \cdot 23.14\ MJ/kg = 0.310\ MJ/100\,kg\ \text{metallic charge},$$

- oxidation of iron:

$0.25 \cdot (1.967 + 0.482)\ kg\ Fe \cdot 4.81\ MJ/kg + 0.005 \cdot (1.967 + 0.482)\ kg\ Fe \cdot 7.35\ MJ/kg = 3.845\ MJ/100\,kg$ metallic charge,

- oxidation of aluminum:

$$0.314\,kg\ Al \cdot 28.52\ MJ/kg = 8.955\ MJ/100\,kg\ \text{metallic charge},$$

- reactions of slag compounds formation:

  a. silicon oxidizing from the melt creates the compound $2 \cdot CaO \cdot SiO_2$ in the slag, as per the exothermic reaction:

$$(CaO) + (SiO_2) = (2 \cdot CaO \cdot SiO_2) \quad (10.21)$$

  where the thermal effect is 2.27 MJ/kg SiO_2. In the example concerned:

  0.45 kg Si · 60/28 · 2.27 MJ/kg = 2.188 MJ/100 kg metallic charge,

  b. phosphorus oxidizing from the melt creates the compound $4 \cdot CaO \cdot P_2O_5$ in the slag, as per the exothermic reaction:

$$4(CaO) + (P_2O_5) = (4 \cdot CaO \cdot P_2O_5) \quad (10.22)$$

  where the thermal effect is 4.73 MJ/kg P_2O_5. In the example concerned:

(0.0066 + 0.0106) kg P · 142/62 · 4.73 MJ/kg = 0.186 MJ/100 kg metallic charge.

The total amount of heat emitted as a result of exothermic reactions is 57.602 MJ/100 kg metallic charge.

4. *The heat from exothermic reactions occurring during the oxidation of graphite electrodes*: It is assumed that, during melting, 0.4 kg electrodes are consumed per 100 kg metallic charge; in addition, 50% of this amount is consumed by oxidation. The electrode graphite oxidizes as per the reaction:

$$C_{gr} + O_2 = \{CO_2\} \quad (10.23)$$

   where the thermal effect is 32.65 MJ/kg C_{gr}. In the example concerned, we obtain:

   0.4 · 50/100 · 32.65 = 6.530 MJ heat/100 kg metallic charge.

5. *The physical heat of charge materials* is calculated from the fundamental law of thermal balance:

$$Q = \sum_{i=1}^{n} m_i c_{wi} (T_i - T_o) \quad (10.24)$$

   where:

   Q – the amount of physical heat from charge materials, KJ,
   n – quantity of charge materials,
   m_i – masses of individual charge materials, kg,

c_{wi} – specific heat of individual charge materials, kJ/kg·K,
T_i – temperature of a charge material i at the moment of putting it into the furnace, K,
T_o – ambient temperature, K.

It can be seen from the formula (10.24), if the charge material temperature at the time of putting it into the furnace is equal to the ambient temperature, the physical heat of this material – calculated for the purposes of the heat balance – is zero. In the example concerned, it is assumed that all of the charge materials used for the steel production are not preheated before use and have the ambient temperature at the time of being put into the furnace. Therefore, the physical heat of charge materials in this case is zero.

The output items in the heat balance are calculated as follows:

1. *Physical heat of steel* is calculated from the formula:

$$Q_{st} = m_{st}\left[c_{wss}\left(T_l - T_o\right) + Q_{ust} + c_{wsc}\left(T - T_l\right)\right] \tag{10.25}$$

where:
Q_{st} – amount of physical heat of steel at a temperature T, KJ,
m_{st} – mass of steel, kg,
c_{wss} – average heat of steel in the solid state, kJ/kg·K,
c_{wsc} – average heat of steel in the liquid state, kJ/kg·K,
Q_{ust} – amount of latent heat of steel melting, kJ/kg,
T – temperature at the end of steelmaking, before tapping, K,
T_l – steel melting temperature, K,
T_o – ambient temperature, K.

We need to know the specific heat of steel in the liquid and solid states for the calculations. It depends on the chemical composition of steel. The exact values of the specific heat can be read from the literature relevant tables. For the needs of the heat balance, we can assume that the average specific heat of low-carbon and medium-carbon steels in the solid state is 0.698 kJ/kg·K, whereas in the liquid state it is 0.836 kJ/kg·K. The latent heat of melting for non-alloyed steels can be assumed as 271.7 kJ/kg. To calculate Q_{st}, we also need to know the melting temperature of the steel produced, which also depends on its chemical composition. We can approximately calculate it from the formula:

$$T_l = T_z - \sum_{i=1}^{n} p_i \cdot k_i \tag{10.26}$$

where:
T_l – steel melting temperature, K,
T_z – iron melting temperature, K,
p_i – percentage share of a specific element in the steel, %,

TABLE 10.16
Loeser Wensel Tables for Computing the Approximate Steel Liquidus Temperature, Depending on the Contents of Its Components [1] (Author's Work)

Element	Content Range (%)	Temperature Reduction Coefficient
Carbon	0.00–0.49	65
	0.50–0.99	70
	1.00–1.99	75
	2.00–2.99	82
	3.00–3.99	94
	4.00–4.99	100
Magnesium	0.00–1.5	5
Silicon	0.00–3.0	8
Phosphorus	0.00–0.7	30
Sulfur	0.00–0.1	25
Copper	0.00–0.3	5
Nickel	0.00–9.0	4
Vanadium	0.00–1.0	2
Molybdenum	0.00–0.3	2
Chromium	0.00–18.0	1.5

k_i – coefficient reducing the steel melting temperature, its value can be read from the tables by Loeser and Wensel (Table 10.16).

In the example concerned, it is assumed that the steel tapping temperature is 1,873 K (1,600°C). The steel melting temperature, calculated from the formula (10.26), is:

$$T_l = 1{,}812 - 0.2 \cdot 65 - 1.11 \cdot 5 - 0.29 \cdot 8 - 0.023 \cdot 30 - 0.016 \cdot 25 = 1{,}790 \text{ K.}$$

For such conditions, the physical heat of steel calculated from the formula (10.25) will be:

$$Q_{st} = 98.005 \cdot [0.698 \cdot (1{,}790 - 293) + 271.7 + 0.836 \cdot (1{,}910 - 1{,}790)] = 138.886$$
MJ/100 kg of metallic charge.

2. *Physical heat of slag* is calculated from the formula:

$$Q_s = m_s \left[c_{ws} (T - T_o) + Q_{us} \right) \tag{10.27}$$

where:

Q_s – amount of physical heat of slag at the temperature T, KJ,
m_s – mass of slag, kg,

c_{ws} – average specific heat of slag, kJ/kg·K,
Q_{us} – amount of latent heat of slag melting, kJ/kg,
T – temperature at the end of melting, before slag tapping, K,
T_o – ambient temperature, K.

The specific heat of slag depends on its chemical composition and temperature. For basic steelmaking in an EAF, the specific heat of the slags applied can be calculated with a sufficient accuracy on the basis of the formula:

$$c_{ws} = 0.736 + 2.93 \cdot 10^{-4} T \tag{10.28}$$

Symbols as in equation (10.27). The latent melting heat of basic slag is approximately 209 kJ/kg [1].

In the example concerned, it is assumed that approximately 50% slag of the meltdown period was tapped, which was 3,431 kg/100kg metallic charge, with the assumed temperature of 1,830 K. The rest of the slag, or 7.183 kg/100kg metallic charge, was tapped after the completion of the melting; its assumed temperature was 1,910 K. Then, the physical heat of the slag is:

$Q_s = 3.343 \cdot [(0.736 + 2.93 \cdot 10^{-4} \cdot 1830 \cdot (1{,}830 - 293) + 209] + 7.183 \cdot [(0.736 + 2.93 \cdot 10^{-4} \cdot 1{,}830 \cdot 1{,}910 \cdot (1{,}910 - 293) + 209] = 23.976$ MJ/100kg metallic charge

3. *Physical heat of gases*: Gases forming during steelmaking in an EAF, determined in the mass balance, are only part of gases that absorb heat in the process. Practically all operating EAFs have the fourth hole in the roof for exhausting gas and dust. Some furnaces also have hoods over the furnace to collect gas and dust from the furnace surroundings. Recently, furnaces are fully encapsulated in special chambers, so-called dog houses. Then, practically all of the gases forming in the manufacturing process, dust, and aspired air are exhausted from the furnace volume above the slag level and from the furnace surroundings to the treatment system.

   Only the heat of the gases escaping through the fourth hole is taken into account in the heat balance. All of the other gases from the furnace surroundings heat up as a result of the radiation and convection of the outer surfaces of the hearth, walls, roof, and the refractory lining when the roof is open. Their heat is included in separately calculated items.

   Computing the physical heat of gases escaping through the fourth hole in the roof can be performed in two ways:

   - it is assumed that gases forming in the manufacturing process (from charge materials, oxidized metal bath constituents and the air aspired from the environment) heat to a temperature of approximately 1,173 K (900°C); their heat is calculated, assuming that the aspired air heats from the process gases,

- the gas and air heat are calculated knowing the capacity of fans exhausting gases through the fourth hole, and by measuring the temperature of the mixture of process gases and air after entering the exhaust system. The estimation of the specific heat of such a mixture of gases poses a problem, as the shares of all its components should be known.

The amount of gases exhausted through the fourth hole in the roof is approximately [6]:

$$I_G = 10^3 \cdot G \tag{10.29}$$

where:

I_G – the amount of gases exhausted through the fourth hole in the roof, Nm3/h,

G – furnace capacity, t.

For instance, for a furnace with a capacity of 80 Mg, the amount of gases exhausted is approximately 80,000 Nm^3/h. For the calculations for a 100 kg metallic charge, the amount of gases exhausted through the fourth hole will be about 100 Nm^3/h.

The average temperature of the gases exhausted, measured at the inlet to the system is 1,773–2,023 K (1,500°C–1,750°C), but at the outlet the system is 473–573 K (200°C–300°C) [7], depending on the furnace design, melting practices applied, steel grades, etc. Moreover, this temperature varies during the subsequent stages of production.

The amount of physical heat contained in the off-gas can be computed using the formula (10.24). Furthermore, m_i in this formula means the masses (or volumes) of individual gases – components of the off-gas, c_{wi} – relevant specific heats.

In the example concerned, it is assumed that, during the metallurgical process, 8.865 Nm^3 gas per hour is emitted per 100 kg metallic charge. The average specific heat of off-gas is assumed as 1.6 $kJ/Nm^3{\cdot}K$. The tap-to-tap time is assumed as 3 h, and the inlet gas temperature as 1,973 K (1,700°C), while the outlet gas temperature as 493 K (220°C). The physical heat of the off-gas is:

$$Q = 8.865 \cdot 3 \cdot 1.6 \cdot (1{,}973 - 493) = 51.36 \text{ MJ/100 kg metallic charge.}$$

4. *Endothermic reaction heat*: During steelmaking in an EAF, endothermic reactions may occur: decomposition of calcium carbonate, calcium hydroxide, passing sulfur from the metal bath to the slag, formation of calcium carbide in the slag, and evaporation of moisture. However, the mentioned reactions occur practically only if non-standard practices are applied (e.g. use of limestone or iron ore in the charge, melting under carbidic slag). For most of the practices applied, the amount of endothermic reaction heat is negligible. In the example concerned, it is also assumed that the amount of heat from endothermic reactions is zero.

5. *Heat losses with the cooling water* are calculated from the basic formula:

$$Q_w = m_w c_w \Delta T \tag{10.30}$$

where:

Q_w – the amount of physical heat of cooling water, KJ,
m_w – mass of the cooling water: equals to the product of density and time and flow rate, kg,
c_w – specific heat of water, kJ/kg·K,
ΔT – temperature difference of cooling water at the outlet and inlet of the cooling system, K.

In modern EAFs, water-cooling is used for the walls and the roof of the furnace and for gas exhaust systems, secondary circuit components, sometimes for electrodes and other minor structural components of the furnace. Heat balance calculations practically only include the amount of heat absorbed by water in the cooling systems of the walls and the roof. The heat amount is calculated according to the formula (10.30).

The amount of cooling water applied in industrial practice depends on the structure of the applied systems and parameters of the cooling water (inlet and outlet temperature, pressure, etc.). The problems related to designing the cooling systems of the EAF are thoroughly described in the literature [1].

For instance, for a 95 metric ton furnace, a cooling system with the following parameters was applied [2]:

- for the wall cooling systems, the cooling water flow rate was 315 Nm^3/h, at the inlet temperature of 302–306 K (29°C–33°C) and the outlet temperature of 307–319 K (34°C–46°C),
- for the roof cooling systems, the cooling water flow rate was 250 Nm^3/h, at the inlet temperature of 302–306 K (29°C–33°C and the outlet temperature of 308–316 K (35°C–43°C).

For a 140 metric ton furnace, a cooling system with the following parameters was applied [2]:

- for the wall cooling systems, the cooling water flow rate was 300 Nm^3/h, at the inlet temperature of 288–293 K (15°C –20°C), and the outlet temperature of 313–323 K (40°C–50°C),
- for the roof cooling systems, the cooling water flow rate was 250 Nm^3/h, at the inlet temperature of 288–293 K (15°C–20°C), and the outlet temperature of 313–323 K (40°C–50°C).

In the example concerned, it is assumed that the melting process is carried out in an EAF with a capacity of 50 t, for which the cooling water flow rate is 40,000 Nm^3/h for the walls and 7,500 Nm^3/h for the roof, respectively, at equal temperatures of water: inlet 293K (20°C) and outlet 308K (35°C) (it is assumed that only part of the roof is cooled with water). Then, the heat losses with the cooling water are:

$Q_w = 40{,}000 \cdot 3 \cdot 4.19 \cdot (308 - 293) + 7{,}500 \cdot 3 \cdot 4.19 \cdot (303 - 293) = 8{,}956.1$ MJ/50 t, as converted per 100 kg metallic charge it is 17.912 MJ of heat.

6. *Heat losses through radiation and convection* occur from the outer surface of the roof, wall shell, and bottom of the furnace during the whole process time from tap to tap, and the inner surface of the roof, walls, bottom, and electrodes during roof opening. The amount of heat lost by radiation can be calculated on the basis of the Stefan-Boltzmann law:

$$Q_r = \varepsilon \cdot C_o \left[\left(\frac{T_1}{100} \right)^4 - \left(\frac{T_2}{100} \right)^4 \right] \cdot F \cdot \tau \qquad (10.31)$$

Q_r – amount of the radiating heat, J,
ε – coefficient of emissivity; for a steel shell of the furnace, it is assumed $\varepsilon = 0.80$,
C_o – constant of black body radiation of 5.67 J/m2·s·K^4,
T_1, T_2 – absolute temperatures of the radiation surface and the environment, respectively, K,
F – radiation surface area, m^2,
τ – radiation time, s.

The amount of heat lost by convection can be calculated on the basis of the formula:

$$Q_c = \alpha \cdot F \cdot (T_1 - T_2) \cdot \tau \qquad (10.32)$$

where:
Q_c – amount of heat carried away by convection J,
α – heat transfer coefficient, W/m^2·K,
other symbols as in equation (10.31).

The value of coefficient α can be calculated as follows:

- for a horizontal plate, top surface:

$$\alpha = 3.3 \cdot \sqrt[4]{T_1 - T_2} \qquad (10.33)$$

- for a horizontal plate, bottom surface:

$$\alpha = 1.6 \cdot \sqrt[4]{T_1 - T_2} \qquad (10.34)$$

- for a vertical plate:

$$\alpha = 2.6 \cdot \sqrt[4]{T_1 - T_2} \qquad (10.35)$$

- for a vertical cylinder:

$$\alpha = 0.72 \cdot \frac{(T_1 - T_2)^{0.23}}{d^{0.3}} \tag{10.36}$$

To use the formulas (10.31) and (10.32), you need to know the areas of the relevant surfaces of individual components of the EAF structure, their temperatures, and the tap-to-tap time and the roof open time. For the dimensions of a specific furnace, it is easy to calculate the areas of individual surfaces. The problems related to designing the dimensions of an EAF are presented thoroughly in the literature [1]. The estimation of the temperatures of individual parts of the furnace requires relevant measurements. Table 10.17 shows an example of average temperatures of the outer parts of an EAF with a capacity of 95 t [2]; the average ambient temperature around the furnace is 300 K (27°C).

In the example concerned, it is assumed that the steal is produced in an EAF with a capacity of 50 t, for which the areas of the relevant furnace surfaces are summarized in Table 10.18. For the assumed temperatures and the relevant surfaces, the amounts of radiation and convection heat were calculated, and they are also included in Table 10.18. The radiation heat and the convection heat were calculated according to formulas (10.31) and (10.32), assuming:

- coefficient of emissivity for the furnace steel shell $\varepsilon = 0.80$,
- coefficient of emissivity for the refractory materials $\varepsilon = 0.80$,
- coefficient of emissivity for the graphite electrodes $\varepsilon = 0.95$,
- tap-to-tap time $\tau = 3\,\text{h}$,
- roof open time $\tau = 10\,\text{min}$,
- time of inner surfaces of the walls and bottom radiation $\tau = 5\,\text{min}$,
- α – as in formulas (10.33)–(10.36),
- ambient temperature 293 K (20°C).

TABLE 10.17
Average Temperatures of the Outer Parts of an Electric Arc Furnace with a Capacity of 95 Mg [2] (Author's Work)

Surface	Surface Area (m²)	Average Temperature during Melting (K)
Water-cooled parts of the walls	36.8	315
Parts of the walls not cooled with water	23.6	511
Water-cooled parts of the roof	28.8	318
Parts of the roof not cooled with water	3.7	523
Bottom	41.4	388

TABLE 10.18
Areas and Average Temperatures of Outer Parts of the EAF and the Amounts of Heat Taken Away by Radiation and Convection in the Example Concerned (Author's Work)

	Area of Surface	Average Temperature during Melting	Amount of Heat Taken Away by Radiation and Convection	
Surface	**m²**	**K**	**MJ**	**MJ**
Outer parts of the walls cooled with water	25	315	30.319	33.448
Outer parts of the walls not cooled with water	16	510	472.503	374.190
Outer parts of the roof cooled with water	9	320	13.737	19.742
Outer parts of the roof not cooled with water	45	520	1,449.370	1,413.131
Bottom, outer surface	51	390	393.861	268.274
Inner surface of the roof	41	1170	1952.591	187.847
Inner surface of the walls and the bottom	51	1220	1,436.573	234.781
Electrodes	21	1300	1,933.430	53.611
Total			7,682.384	2,585.024

The total heat losses resulting from radiation and convection are 10,267.408 MJ per a furnace of 50 t. Translating this value into a 100 kg metallic charge, 20.535 MJ heat loss is obtained.

7. *Electric loss* arising from the emission of heat as a result of the current flow through a conductor can be established on the basis of Joule-Lenz law. In a simplified form, this law can be presented as follows:

$$Q_e = 3 \cdot I_m^2 \cdot R_z \cdot \tau \tag{10.37}$$

where:

Q_e – the amount of heat emitted as a result of the current flow through a conductor, J,
I_m – current intensity, A,
R_z – equivalent resistance of a single phase of the furnace power system,
τ – tap-to-tap time, s.

Each EAF is characterized by a specific equivalent resistance of its power supply system. It can be calculated on the basis of the analysis of the furnace electrical supply system or can be obtained based on direct measurements of a specific furnace.

In the example concerned, it is assumed that, for a 50 t furnace equipped with a 30 MVA transformer, the computed equivalent resistance of a single phase of the power supply system is 0.35 mΩ. It is assumed that the average current intensity during the steelmaking is 30 kA. Then, the amount of electric loss is:

$$Q_e = 3 \cdot 30{,}000^2 \cdot 0.35 \cdot 10^{-3} \cdot 3 \cdot 3{,}600 = 10{,}206 \text{ MJ}.$$

Translating this amount into a 100 kg metallic charge, we obtain 20.412 MJ per heat.

The final summary of the heat balance of the heat is shown in Table 10.19.

The calculated electricity consumption for the data assumed in this example is 208.929 MJ/100 kg metallic charge. It corresponds to the energy consumption of 580.4 kWh/Mg metallic charge.

The presented method of computing the heat balance of steelmaking in an EAF assumes that all of the major balance items are determined, apart from the electricity consumption, which closes the balance. In practice, the electricity consumption is measured for each furnace and each heat with high accuracy, and the measurement is easy to make. Therefore, the established amount of electrical energy from the calculations for a specific heat is compared with the measured value. The resulting differences indicate the calculation accuracy. Due to the measuring difficulties of some parameters, being the basis for calculations of individual components of the heat balance, very often in practical calculations the balance is closed with another parameter, e.g. heat loss with escaping gases as well as radiation and convection loss.

The determination of the mass balance and the heat balance of the steelmaking in an EAF using classic computing methods is rather laborious. The application of computer technology can facilitate the computing, in particular when calculations are repeated for lots of steel heats produced with the same practice.

TABLE 10.19
Heat Balance of Steelmaking in an Electric Arc Furnace (Author's Work)

Inputs	MJ	%	Output	MJ	%
1. Electricity	208.929	76.52	1. Steel	138.866	50.86
2. Heat from burners	0.0	0.0	2. Slag	23.970	8.78
3. Exothermic reactions	57.602	21.09	3. Off-gas	51.360	18.81
4. Electrode oxidation	6.530	2.39	4. Endothermic reactions	0.0	0.0
5. Charge materials	0.0	0.0	5. Cooling water loss	17.912	6.56
			6. Radiation & convection	20.535	7.52
			7. Electric loss	20.412	7.47
Total	273.061	100	Total	273.061	100

The practical development of a computer program, including the possibility of computing the mass balance and heat balance for all the varieties of steelmaking practice in EAFs applied, is difficult to accomplish and pointless. From the practical point of view, it is sufficient to develop separate programs for the individual types of practices applied (e.g. with the complete oxidation period, recovery, only meltdown for secondary treatment).

REFERENCES

1. Karbowniczek M.: *Elektrometalurgia stali. Cwiczenia*, Wydawnictwa AGH, Krakow, 1993 (in Polish).
2. Borroni A., Joppolo C., Mazza B, Nano G.: Energy balance measurements for an electric arc steelmaking furnace, *VII UIE Congress*, 1984, Stockholm, Art No. 2.2.7.
3. Reichell W., Zangs L.: Energietckgewinnung aus Lichtbogenofen, *Elektrowarme International*, Vol. 36, 1978, No. B5, pp. B227–B282 (in German).
4. Klein K., Paul G.: Reflections on the Possibilities and Limitations of Cost Saving in Electric Arc Furnaces Steel Production, *Ironmaking and Steelmaking*, Vol. 16, 1989, No. 1, pp. 25–34.
5. Fettwels W., Nagla R., Schunk E.: Operations of Electric Arc Furnace Using Computer Control, *Iron and Steel Engineer*, Vol. 62, 1985, No. 8, pp. 68–72.
6. Raguin J., Vizios J.: Les depoussierages an accierie electrique, *Circulare d'Informations Techniques du CDS*, Vol. 12, 1975, pp. 2021–2641.
7. Flux J.H.: The clean melting shop, *Ironmaking and Steelmaking*, Vol. 3, 1976, No. 4, pp. 208–214.

Index

For Product Safety Concerns and Information please contact our EU representative GPSR@taylorandfrancis.com
Taylor & Francis Verlag GmbH, Kaufingerstraße 24, 80331 München, Germany

www.ingramcontent.com/pod-product-compliance
Lightning Source LLC
Chambersburg PA
CBHW060353220326
41598CB00023B/2906

* 9 7 8 0 3 6 7 6 7 3 4 8 2 *